PHARMACIST'S GUIDE TO LIPID MANAGEMENT

PHARMACIST'S GUIDE TO LIPID MANAGEMENT

Edited by
Barbara S. Wiggins, Pharm.D., BCPS (AQ Cardiology), CLS, FAHA
Joseph J. Saseen, Pharm.D., FCCP, BCPS (AQ Cardiology), CLS
Sarah A. Spinler, Pharm.D., FCCP, BCPS (AQ Cardiology)

AMERICAN COLLEGE OF CLINICAL PHARMACY
LENEXA, KANSAS

American College of Clinical Pharmacy
13000 W. 87th Street Parkway
Lenexa, KS 66215-4530
Telephone: (913) 492-3311
Fax: (913) 492-0088
www.accp.com
E-mail: accp@accp.com

Director of Professional Development: Nancy M. Perrin, M.A., CAE
Publications Project Manager: Janel Mosley
Medical Copy Editor: Kimma Sheldon, Ph.D., M.A.
Desktop Publisher/Graphic Designer: Jen DeYoe, B.F.A.

Editors
Barbara S. Wiggins, Pharm.D., BCPS (AQ Cardiology), CLS, FAHA
Joseph J. Saseen, Pharm.D., FCCP, BCPS (AQ Cardiology), CLS
Sarah A. Spinler, Pharm.D., FCCP, BCPS, CLS

Copyright © 2008 by the American College of Clinical Pharmacy. No part of this publication may be reproduced, stored in a retrieval system, or transmitted, in any form or by any means, electronic or mechanical, including photocopy, without prior written permission of the American College of Clinical Pharmacy.

Printed in the United States of America.

Library of Congress Control Number: 2008933605
ISBN: 978-1-932658-63-7

The American College of Clinical Pharmacy, the editors, and the authors would like to thank the following individuals for their careful review of this text.

Krystal L. Edwards, Pharm.D., BCPS
Assistant Professor
Texas Tech University HSC School of Pharmacy
Dallas, Texas

Eric K. Gupta, Pharm.D., BCPS
Assistant Professor of Pharmacy Practice
Western University of Health Sciences
Pomona, California

Daniel E. Hilleman, Pharm.D., FCCP
Professor of Pharmacy Practice
Creighton University School of Pharmacy
Creighton Cardiac Center
Omaha, Nebraska

Donald Lamprecht, Pharm.D., BCPS
Clinical Pharmacy Specialist
Cardiac Risk Service
Kaiser Permanente, Colorado
Aurora, Colorado

Eric J. MacLaughlin, Pharm.D., BCPS
Associate Professor and Division Head of Adult Medicine
Department of Pharmacy Practice
Texas Tech University Health Sciences Center School of Pharmacy
Amarillo, Texas

Robert Lee Page II, Pharm.D., FCCP, BCPS (AQ Cardiology)
Associate Professor of Clinical Pharmacy and Physical Medicine
Clinical Specialist, Division of Cardiology and Heart Transplant
University of Colorado Denver Schools of Pharmacy and Medicine
Aurora, Colorado

Toby C. Trujillo, Pharm.D., BCPS
Associate Professor of Clinical Pharmacy
University of Colorado Denver School of Pharmacy
Clinical Coordinator, University of Colorado Hospital
Aurora, Colorado

DISCLOSURE OF POTENTIAL CONFLICTS OF INTEREST

Consultancies: Kim K. Birtcher (Tarascon); Matthew K. Ito (CV Therapeutics); Carol Mason (AstraZeneca; Abbott Laboratories); Joseph J. Saseen (AstraZeneca); Sarah A. Spinler (Sanofi-Aventis)

Grants: Michael B. Bottorff (Bristol-Myers Squibb); Matthew K. Ito (Merck/Schering-Plough); Sarah A. Spinler (AstraZeneca)

Honoraria: Michael B. Bottorff (AstraZeneca, speaker; Sanofi-Aventis, speaker; Pfizer, speaker); Eric K. Gupta (AstraZeneca, speaker; Pfizer, speaker); Daniel Hillerman (AstraZeneca, speaker; Pfizer, speaker; Abbott Laboratories, speaker); Matthew K. Ito (AstraZeneca, speaker; Merck/Schering-Plough, speaker); Ralph LaForge (Abbott Laboratories, speaker; AstraZeneca, speaker; Pfizer, speaker; Takeda Pharmaceuticals, speaker); Carol Mason (Pfizer, speaker; Merck/Schering-Plough, speaker; Kos Pharmaceuticals, speaker); Joseph J. Saseen (Daiichi Sankyo Pharma, speaker); Evan M. Sisson (Novartis, speaker); Sarah A. Spinler (ASHP Advantage, speaker; Bristol-Myers Squibb, speaker; Sanofi-Aventis, speaker; GlaxoSmithKline, speaker); Barbara M. Wiggins (Pfizer, speaker)

CONTENTS

FOREWORD vii

1 PATHOPHYSIOLOGY OF ATHEROSCLEROTIC VASCULAR DISEASE 1

Role of Cholesterol and Lipids 4
Lipid Metabolism and Transport 5
Role of Genetics 10
Familial Dyslipidemias 10
Secondary Lipid Disorders 14

2 CLINICAL EVALUATION OF THE ADULT PATIENT 20

Identify Dyslipidemias 20
Emerging Risk Markers 26

3 REDUCING CORONARY HEART DISEASE RISK THROUGH LIFESTYLE MODIFICATION 30

Reducing CHD Risk with Nutritional Therapy Recommendations 31
Plant Stanols/Sterols 34
Dietary Fiber 35
Exercise 36
Smoking Cessation 37
Weight Loss 38
Summary 39

4 METABOLIC SYNDROME 42

Prevalence of Cardiometabolic Risk Factors in the United States 43
The Pathophysiology of Metabolic Syndrome 44
Identification and Management of Metabolic Syndrome in Patients 49
Conclusion 55

5 PHARMACOTHERAPY 60

Statins 60
Cholesterol Absorption Inhibitors 67
Bile Acid Sequestrants 71
Fibric Acid Derivatives 73
Nicotinic Acid 76
Omega-3 Fatty Acids 79

6 LANDMARK CLINICAL TRIALS AND OTHER RELEVANT PUBLICATIONS ... 84

Clinical Guidelines and Expert Consensus Documents ... 84
Clinical Trials Evaluation: Cardiovascular End Points ... 85
Clinical Trials Assessing Surrogate Markers of Atherosclerosis ... 94
Appendix: Clinical Trial Acronym Glossary ... 99

7 SPECIAL PATIENT POPULATIONS ... 100

Pregnancy ... 100
Children ... 103
Acute Coronary Syndrome Background ... 106
Chronic Kidney Disease Background ... 109
Human Immunodeficiency Virus ... 111
Solid Organ Transplantation ... 120
Elderly ... 122

8 CLINICAL PRACTICE PEARLS ... 139

Introduction ... 139
Novel and Emerging Risk Factors ... 139
HMG-CoA Reductase Inhibitors (Statins) ... 144
Fibric Acid Derivatives (Fibrates) ... 147
Intestinal-Acting Agents ... 147
Niacin ... 150
Nonprescription Products ... 152

9 SYSTEMATIC MANAGEMENT OF LIPID DISORDERS IN PHARMACEUTICAL CARE ... 156

Appropriate Entry Criteria and Sufficient Patient Referral for Specialized Lipid Clinic Services ... 158
Patient Evaluation ... 158
Patient Education ... 158
Physical Assessment ... 160
Laboratory Values ... 161
Adherence ... 162
The Pharmaceutical Care Treatment Plan ... 163
Practice Models ... 164
Summary ... 165

INDEX ... 169
CONTRIBUTORS ... 191

FOREWORD

It's not knowing what to do, it's doing what you know.
                                    Anthony Robbins (motivational speaker)

Knowledge and opportunity go hand in hand. Nowhere is this better illustrated than in the nationwide effort to reduce the risk of cardiovascular disease (CVD), America's number one killer. Research has clearly laid the blame for CVD on atherosclerosis, and at the nexus of its development is dyslipidemia. Research has also provided us with a powerful group of drugs to effectively and safely manage dyslipidemia—including statins, most notably—and abundant evidence of a direct relationship between low-density lipoprotein cholesterol (LDL-C) and CVD risk; the lower the on-treatment LDL-cholesterol value, the lower the CVD risk.

However, knowing this by itself is not going to help anyone; action is required. And action in this case provides health professionals an opportunity to make a significant difference, both in prolonging people's lives (reducing premature death) and in improving their quality of life (reducing CVD events). For all who seize this opportunity, this guide is a great resource. Whether you are a student learning the fundamentals of lipid management, a new practitioner developing a lipid management service, or an accomplished clinician needing a quick fact or reference, this book is for you. I commend the authors and editors for putting such a complete, authoritative, and clinically relevant reference in our hands.

Patients at risk of CVD cry out for our help. They need assistance finding effective regimens that will give them maximal benefit. They also want safe regimens and reassurance of their safety. Furthermore, they need help learning about the therapy; their adherence will ensure that they derive the greatest benefit. These services constitute the practice of clinical lipidology, which was recently recognized as a medical subspecialty. In fact, boards are now available to physicians, pharmacists, and other health professionals to certify the acquisition of special skills and knowledge for the practice of clinical lipidology. (See www.lipid.org for further information.)

So, with knowledge, here is our opportunity. Seize it. Make a difference.

                                    James M. McKenney, Pharm.D.
              Professor Emeritus, Virginia Commonwealth University
                President and CEO, National Clinical Research, Inc.
  Member, Adult Treatment Panels II and III, National Cholesterol Education
                          Program, National Institutes of Health

PATHOPHYSIOLOGY OF ATHEROSCLEROTIC VASCULAR DISEASE

Matthew K. Ito, Pharm.D., FCCP, CLS
Professor and Chair of Pharmacy Practice

Joel C. Marrs, Pharm.D., BCPS
Clinical Assistant Professor of Pharmacy Practice

Atherosclerotic vascular disease is a progressive process that begins in the first decade of life; it is considered a chronic low-grade inflammatory process in response to injury of the vascular endothelium.[1] Factors such as glycoxidation products associated with diabetes, inflammatory cytokines and free fatty acids produced by adipocytes, neurohormonal abnormalities, bacterial products, shear stress, and lipotoxicity can cause vascular injury. The damaged vascular endothelium becomes prothrombotic and chemotactic to inflammatory cells and platelets, with attenuated endothelial-dependent vasodilation. Lipid accumulation beneath the endothelium leads to the formation of a fatty streak, increased extracellular lipids, and abluminal growth to form an early atheroma. Further growth and immune cell activation lead to atheromas that are necrotic and prone to rupture or erosion, leading to thrombosis. Plaque ruptures are found beneath about 60–70% of the thrombi responsible for acute coronary syndromes.[2]

In patients with elevated serum cholesterol concentrations, excess apolipoprotein B–containing lipoproteins (very low-density lipoprotein [VLDL] remnant, intermediate-density lipoprotein [IDL], low-density lipoprotein [LDL], and lipoprotein A [Lp(a)]) migrate between the endothelial junction and into the subendothelial space, particularly at sites of hemodynamic stress (Figure 1). The infiltration of lipoproteins into the artery wall and their subsequent retention (by binding to proteoglycans) initiate an inflammatory response.

FIGURE 1. THE PROCESS OF ATHEROGENESIS

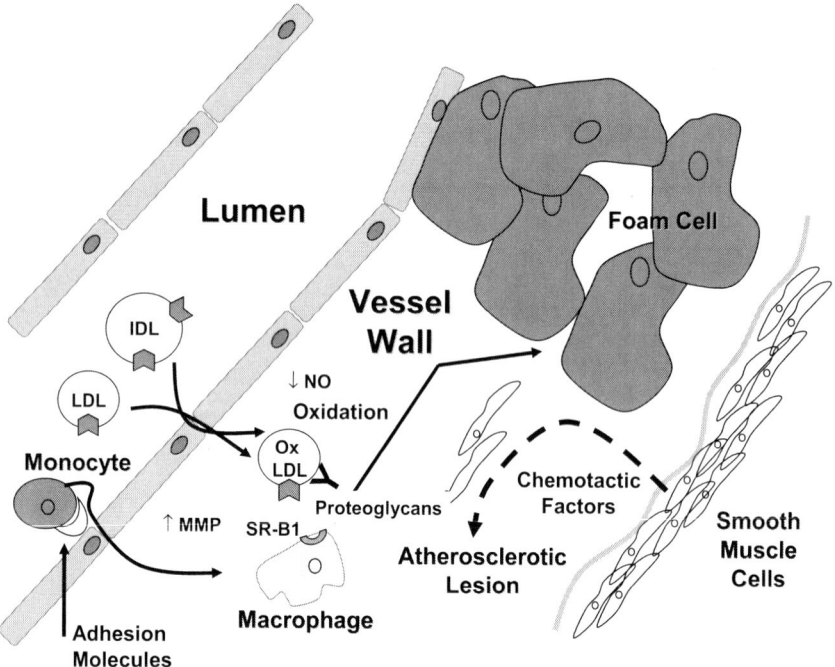

MMP = matrix metalloproteinases; NO = nitric oxide.

Atherosclerosis is initiated by the migration of LDL and remnant lipoprotein particles into the vessel wall. These particles undergo oxidation and are taken up by macrophages in an unregulated fashion. The oxidized particles induce endothelial cell dysfunction, which leads to a reduced ability of the endothelium to dilate the artery and cause a prothrombotic state. The unregulated uptake of cholesterol by macrophages leads to foam cell formation and the development of a blood clot favoring a fatty lipid core. The enlarged lipid core eventually causes an encroachment of the vessel lumen. Early in the process, smooth muscle cells are activated and recruited from the media to the intima, helping produce a collagen matrix that covers the growing clot favoring the lipid core, protecting it from circulating blood. Later, macrophages produce and secrete MMPs that degrade the collagen matrix, leading to an unstable plaque that could cause a myocardial infarction. (Modified from Ito MK. Hyperlipidemia. In: Chisholm MA, Schwinghammer TL, Wells BG, DiPiro JT, Kolesar JM, Malone PM, eds. Pharmacotherapy: Principles & Practice. New York: McGraw-Hill, 2007:chap. 9; with permission.)

The greater density of proteoglycans in the subendothelial space (e.g., bifurcation) increases lipoprotein retention.[3] After entering the intima, lipoproteins are then structurally modified by oxidation. Small, denser LDL particles migrate into the arterial wall more readily and are particularly susceptible to oxidation. Phospholipids released from oxidized lipoproteins[4] as well as other cytotoxic agents promote endothelial dysfunction by disturbing the production of vasoactive molecules such as nitric oxide (NO). Endothelium-derived NO, synthesized from endothelial NO synthase from the precursor L-arginine, maintains endothelium-dependent vasodilation and a balance between profibrinolytic and prothrombotic activity. Reductions in NO lead to the overexpression of plasminogen activator inhibitor-1 (PAI-1), which is the major physiologic inhibitor of tissue-type plasminogen activator (t-PA). Under normal physiologic states, endogenous t-PA is neutralized by PAI-1, which binds to t-PA and forms a stable 1:1 stoichiometric complex, maintaining net fibrinolytic activity. Elevated PAI-1 concentrations appear to increase the risk of atherothrombosis.[5]

Other mediators that modulate vascular tone and platelet aggregation are the prostanoids, such as prostacyclin (PGI_2) and thromboxane (TXA_2), respectively. Prostanoids are produced from a variety of cells such as vascular endothelial cells, platelets, and inflammatory cells from arachidonic acid by the enzyme cyclooxygenase 1 and 2 (COX-1 and COX-2) and specific synthases (PGI_2 synthase and TXA_2 synthase). Prostanoids have been implicated in a variety of physiologic processes in atherosclerosis and thrombosis. A normal physiologic balance in the various prostanoids and enzymes (COX-1 and COX-2) is important in maintaining a healthy vascular environment. Endothelial injury and recruitment of inflammatory cells in the arterial wall can upset this balance between PGI_2 and TXA_2 and promote thrombosis.

Dysfunctional endothelial cells release chemotactic factors such as monocyte chemotactic protein-1 and macrophage colony stimulating factor that recruit circulating monocytes and T lymphocytes to the endothelium. They also express cell adhesion molecules such as intracellular adhesion molecule-1, vascular cell adhesion molecule-1, and E-selectin, which cause the recruited inflammatory cells rolling along the endothelial surface to adhere at the site of activation. Healthy endothelial cells do not interact with blood cells because they lack surface adhesion molecules.[6] Chemokines produced in the intima stimulate the transmigration of monocytes and lymphocytes into the intima. The monocytes induced by growth factors differentiate into macrophages and express scavenger receptors, allowing enhanced uptake of oxidized apolipoprotein B lipoproteins. A second receptor expressed by macrophages is the toll-like receptor. Toll-like receptors recognize components and products of microorganisms and play a vital role in innate immunity.[7] Microbial pathogens have been found in atherosclerotic lesions and are causally implicated in atherogenesis.[8] As the macrophages continue to accumulate lipoproteins, they ultimately develop into lipid-laden foam cells and eventually a fatty streak. At this point, the inflammatory process has become chronic, and further accumulation of foam cells leads to the formation of a lipid-rich core, which marks the transition to a more complicated atherosclerotic plaque.

Vascular wall remodeling leading to outward growth of the wall occurs to accommodate this lipid-rich core. Thus, at this stage of plaque development, the vascular lumen is relatively well preserved and generally cannot be detected by traditional coronary angiographic techniques. Initially, smooth muscle cells migrate and proliferate from the media to the intima stimulated by various biochemical factors, extracellular matrix components, and physical factors such as stretch and shear stress, forming a protective fibrous cap containing collagen and elastin, which separates the potentially thrombogenic lipid core from circulating blood. The thickness and collagen content of the fibrous cap are important variables in plaque stability. As the plaque matures, inflammatory cells secrete matrix metalloproteinases (MMPs) that degrade the collagen and fibrin produced by the smooth muscle cells, leading to a weakened fibrous cap.

Oxidized lipoproteins activate the nuclear factor kappa B expression of MMP-2 and

MMP-9 by macrophages. Nuclear factor kappa B appears to play a role in the regulation of inflammation, cell proliferation, and apoptosis or programmed cell death.[9] Throughout all stages of atherogenesis, macrophages undergo apoptosis. Phagocytic clearance of apoptotic cells is more proficient in early than in advanced lesions. The increase in necrotic debris promotes further inflammation, enhancing plaque instability.[10] Thrombosis and ischemic events result when the fibrous cap tears. These unstable lesions usually outnumber the more stable plaques, thus accounting for a majority of acute coronary syndromes. Evidence demonstrates that aggressive lipid lowering stabilizes these vulnerable lesions and restores endothelial function.[1,11]

In contrast, repeated wound healing secondary to less significant plaque disruption (erosion) that causes no symptoms might produce a more stable plaque because of the smooth muscle cell, collagen, and fibrin accumulation and a resolution of the lipid core.[6] These more stable plaques usually cause luminal encroachment (detected by traditional coronary angiographic techniques) and produce angina pectoris.

ROLE OF CHOLESTEROL AND LIPIDS

Cholesterol is an essential substance manufactured by most cells in the body. Cholesterol is used for maintaining cell wall integrity and for the biosynthesis of bile acids and steroid hormones.[12] The biosynthesis of cholesterol in humans follows a circadian rhythm. Cholesterol synthesis is lower at noon and highest at midnight. However, delayed mealtimes can slow the minimum and maximum cholesterol synthesis rates.[13] The amount of cholesterol in the body is determined by dietary and biliary cholesterol absorption, primarily by the de novo synthesis of cholesterol in the liver and extrahepatic tissue. About 800 mg of cholesterol is synthesized daily, and 300 mg is derived from the diet. To maintain cholesterol homeostasis, the combined amount (1100 mg) must be excreted daily as fecal sterols.[14] Cholesterol is synthesized beginning with three molecules of acetyl-CoA to form 3-hydroxy-3-methyl glutaryl coenzyme A (HMG-CoA). Then, HMG-CoA reacts with HMG-CoA reductase, the enzyme that is competitively inhibited by HMG-CoA reductase inhibitors or statins, together with nicotinamide adenine dinucleotide phosphate, to form mevalonate. Several cholesterol precursors and cholesterol are eventually formed by a series of biochemical reactions. Sterol precursors such as geranyl pyrophosphate and farnesyl pyrophosphate are important modulators of the small proteins involved in cell signaling. The reduction in inflammation, improved vasodilation, and reduced thrombogenicity by the statins may be related to the reduction in sterol precursors.[15]

Other major lipids in the body are triglycerides and phospholipids. Triglycerides are an important source of stored energy in adipose tissue. Triglycerides are synthesized from three molecules of fatty acids esterified to glycerol. Fatty acids are produced in the liver, intestine, and adipose tissue from glucose—a reaction that is regulated by glucagons, insulin, catecholamines, and somatostatin. Patients with prediabetes and diabetes commonly have elevated triglyceride concentrations secondary to increased intra-abdominal adiposity and lipolysis. Phospholipids are a class of lipids formed from fatty acids, a negatively charged phosphate group, nitrogen-containing alcohol, and a glycerol backbone. Phospholipids are essential for cellular function and the

TABLE 1. PHYSICAL CHARACTERISTICS OF LIPOPROTEINS

Lipoprotein	Density Range (g/mL)	Size (nm)	Composition (%) Cholesterol	Composition (%) Triglycerides	Apolipoprotein
Chylomicrons	< 0.95	100–1000	3–7	85–95	A-I, A-II, A-IV, B-48, C-I, C-II, E
VLDL	< 1.006	40–50	20–30	50–65	B-100, C-I, C-II, C-III, E
IDL	1.006–1.019	25–30	40	20	B-100, E
LDL	1.019–1.063	20–25	51–58	4–8	B-100
Lp(a)	1.05–1.12	—	—	—	B-100, apoA
HDL	1.063–1.21	6–10	18–25	2–7	A-I, A-II, C-I, C-II, C-III, E

HDL = high-density lipoprotein; IDL = intermediate-density lipoprotein; LDL = low-density lipoprotein; Lp(a) = lipoprotein A; VLDL = very low-density lipoprotein.

transport of lipids in the circulation. The hydrophilic heads and the hydrophobic tails of the phospholipids form a membrane bilayer where the hydrophobic tails line up against each other. This allows the phospholipids to form the membrane bilayer of lipoproteins spontaneously.

LIPID METABOLISM AND TRANSPORT

Because cholesterol is a relatively water-insoluble molecule, it is unable to circulate through the blood alone. Cholesterol, together with triglycerides and phospholipids, is packaged in a large carrier protein called a lipoprotein (Figure 2). Lipoproteins are water soluble, which allows transportation of the major lipids in the blood. These lipoproteins are spherical and vary in size (about 6–1000 nm) and density (less than 0.94–1.210 g/mL),[12] as listed in Table 1. The amount of cholesterol, triglycerides, and protein varies by lipoprotein size. The major lipoproteins in descending size and ascending density are chylomicrons, VLDL, IDL, LDL, and high-density lipoprotein (HDL). When clinical laboratories measure and report serum total cholesterol, they are measuring the total cholesterol concentration in all the major lipoproteins. In addition, LDL cholesterol (LDL-C) is an estimated value using the Friedewald equation:

$$LDL\text{-}C = total\ cholesterol - (HDL\text{-}C + triglycerides/5).$$

If serum triglyceride concentrations are greater than 400 mg/dL, this formula becomes inaccurate, and LDL-C must be directly measured (discussed in later chapters).[16]

Each lipoprotein has various proteins called apolipoproteins embedded on the surface. These apolipoproteins serve four main purposes. They are required for the assembly and secretion of lipoproteins (such as apolipoprotein B-48 and B-100); they also serve as major structural components of lipoproteins and can act as ligands (apolipoprotein B-100 and apolipoprotein E) for binding to receptors on cell surfaces

FIGURE 2. LIPOPROTEIN STRUCTURE

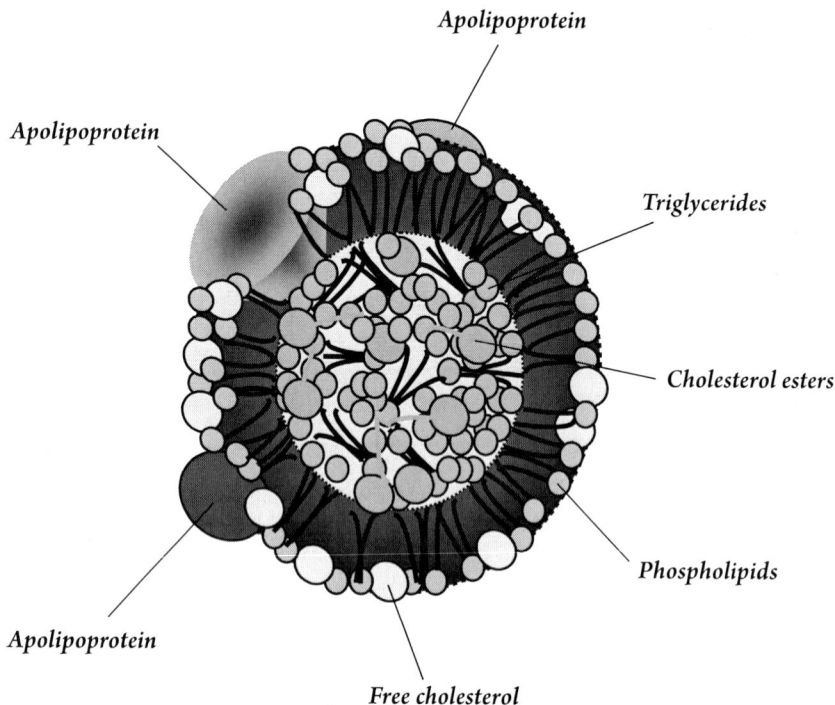

Lipoproteins are a diverse group of particles with varying size and density. They contain various amounts of core cholesterol esters and triglycerides and have varying numbers and types of surface apolipoproteins. The apolipoproteins direct the processing and removal of individual lipoprotein particles. (From Ito MK. Hyperlipidemia. In: Chisholm MA, Schwinghammer TL, Wells BG, DiPiro JT, Kolesar JM, Malone PM, eds. Pharmacotherapy: Principles & Practice. New York: McGraw-Hill, 2007:chap. 9; with permission.)

(LDL or apolipoprotein B/E receptors). In addition, they can be cofactors (such as apolipoprotein CII) for the activation of enzymes (such as lipoprotein lipase or LPL) involved in the breakdown of triglycerides from chylomicrons and VLDL. Apolipoproteins A-I and A-II are major structural proteins on the surface of HDL. Apolipoprotein A-I interacts with adenosine triphosphate–binding cassette (ABC) A1 (ABCA1) and G1 (ABCG1) to traffic cholesterol from extrahepatic tissue (such as the arterial wall) to HDL.[12]

EXOGENOUS LIPID TRANSPORT

Cholesterol from the diet as well as from bile enters the small intestine, where it is emulsified by bile salts into micelles (Figure 3). These micelles interact with the duodenal and jejunal enterocytes' brush border surface. Cholesterol and other sterols are transported from the micelles into these cells by the Niemann-Pick C1-Like 1 transporter.[17] Some cholesterol and most plant sterols, which are structurally similar to cholesterol, are exported back from the enterocyte into the intestinal lumen by the ABC G5/G8 transporter. Cholesterol within enterocytes is esterified by acyl coenzyme acyltransferase and packaged into chylomicrons together with triglycerides, phospholipids, and apolipoprotein B-48, which are then released into the lymphatic circulation. In the circulation, chylomicrons acquire apolipoproteins C and E from HDL; they are converted to chylomicron remnants (through the loss of triglycerides by the interaction of apolipoprotein C-II and LPL). As chylomicrons get smaller, surface lipids and proteins are transferred between HDL particles. Particles of chylomicron remnants are then taken up by endocytosis in the liver by LDL-related protein (LRP).[12]

FIGURE 3. INTESTINAL CHOLESTEROL ABSORPTION AND TRANSPORT

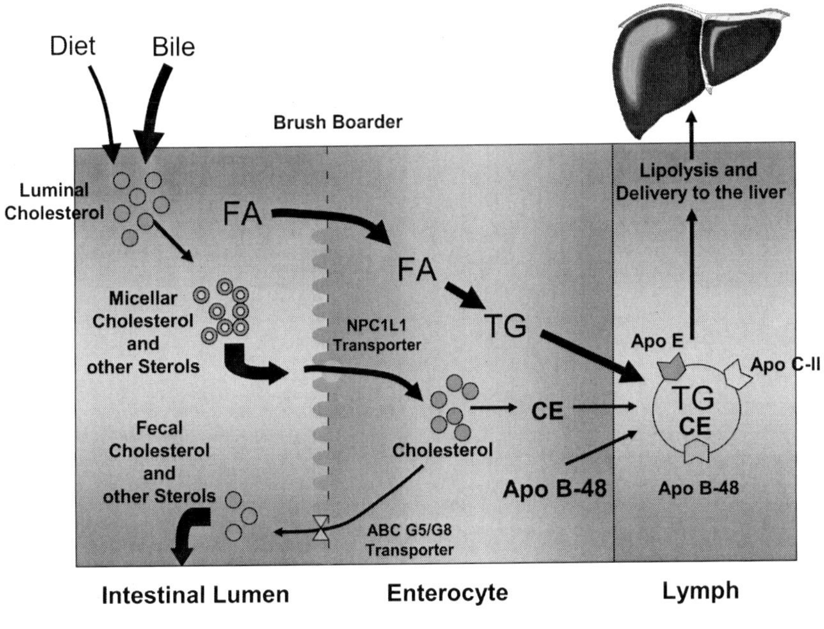

FA = fatty acid; NPC1L1 = Niemann-Pick C1 Like 1; TG = triglycerides; CE = cholesterol ester; ABC G5/G8 = ATP-binding cassette G5/G8

Cholesterol from the food and the bile enters the gut lumen and is emulsified by bile acids into micelles. Micelles bind to the intestinal enterocytes, and cholesterol and other sterols are transported from the micelles into the enterocytes by a sterol transporter. Triglycerides synthesized by absorbed fatty acids, together with cholesterol and apolipoprotein B-48, are incorporated in the chylomicrons. Chylomicrons are released into the lymphatic circulation, acquire apolipoproteins C and E from HDL, and are converted to chylomicron remnants (through the loss of triglycerides). The hepatic LRP then takes them up. (From Ito MK. Hyperlipidemia. In: Chisholm MA, Schwinghammer TL, Wells BG, DiPiro JT, Kolesar JM, Malone PM, eds. Pharmacotherapy: Principles & Practice. New York: McGraw-Hill, 2007:chap. 9; with permission.)

FIGURE 4. ENDOGENOUS LIPOPROTEIN METABOLISM

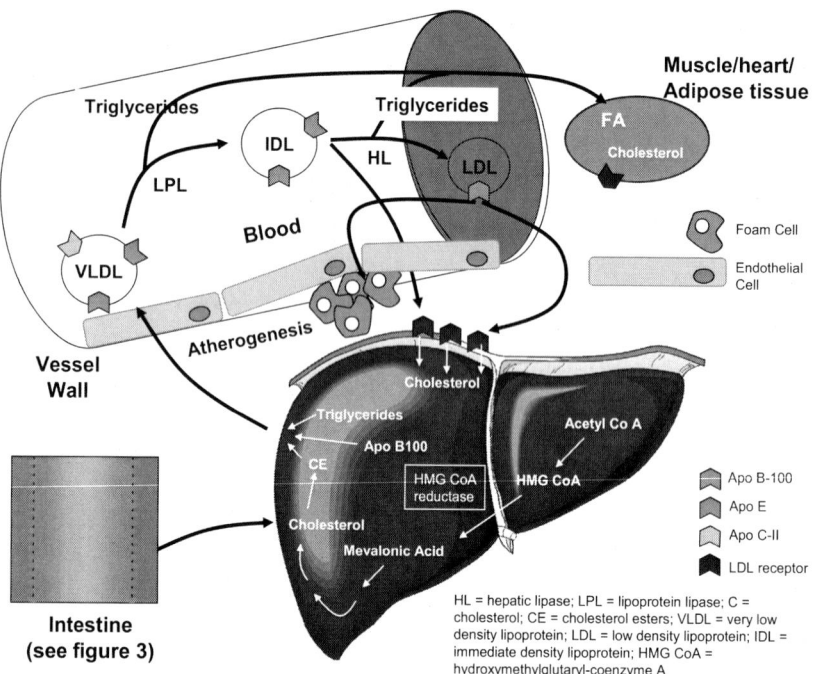

In liver cells, cholesterol and triglycerides are packaged into VLDL particles and exported into the blood, where they are converted to IDL. Intermediate-density lipoproteins can be cleared by hepatic LDL receptors or further metabolized to LDL. LDL can be cleared by the hepatic LDL receptors or can enter the arterial wall, contributing to atherosclerosis. (From Ito MK. Hyperlipidemia. In: Chisholm MA, Schwinghammer TL, Wells BG, DiPiro JT, Kolesar JM, Malone PM, eds. Pharmacotherapy: Principles & Practice. New York: McGraw-Hill, 2007:chap. 9; with permission.)

ENDOGENOUS LIPID TRANSPORT

In the liver, cholesterol and triglycerides are incorporated in VLDL together with phospholipids and apolipoprotein B-100 (Figure 4). Very LDL particles are released into the circulation, where they acquire apolipoproteins E and C-II from HDL. Very LDL loses its triglyceride content through the interaction with LPL to form VLDL remnants and IDL. Intermediate-density lipoprotein can be cleared from the circulation by hepatic LDL receptors or converted to LDL (by further depletion of triglycerides) through the action of hepatic lipases (HLs). About 50% of IDL is converted to LDL. Low-density lipoprotein particles are cleared from the circulation primarily by hepatic LDL receptors by interaction with apolipoprotein B-100, or they can be taken up by extrahepatic tissues or enter the arterial wall, contributing to atherogenesis.[1,12]

REVERSE CHOLESTEROL TRANSPORT

Cholesterol and phospholipids are transported from the arterial wall or other extrahepatic tissues back to the liver by HDL (Figure 5). Cholesterol returned to the liver can be excreted into the bile. This process is referred to as reverse cholesterol transport. Poorly lipidated apolipoprotein A-I (derived from the intestine and liver) on nascent or discoidal HDL interacts with the ABCA1 transporter on extrahepatic tissue. Cholesterol in nascent HDL is esterified by lecithin cholesterol acyltransferase (LCAT), which is activated by apolipoprotein A-1, resulting in a spherical HDL3 particle. Phospholipid transfer protein facilitates the transfer of phospholipids from apolipoprotein B–containing particles to HDL. High-density lipoprotein 3 particles can acquire additional free cholesterol from extrahepatic tissue by interacting with ABCG1, generating HDL2 particles. The esterified cholesterol can be transferred, as noted previously, to apolipoprotein B–containing particles in exchange for triglycerides through the action of cholesterol ester transfer protein (CETP). Triglyceride-rich HDL is hydrolyzed by HL, generating fatty acids and nascent HDL particles, or the mature HDL2 particle can bind to the scavenger receptors (SR-BI) on the hepatocytes and transfer their cholesterol ester content.[12]

FIGURE 5. REVERSE CHOLESTEROL TRANSPORT

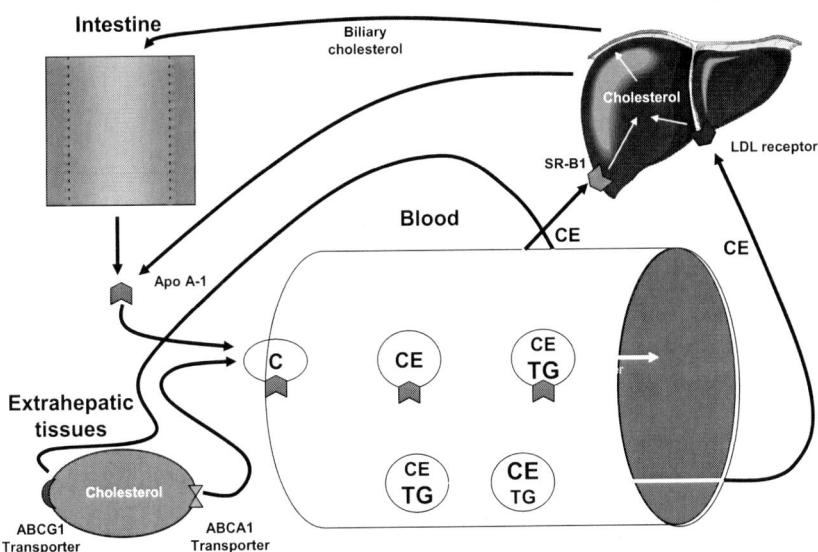

CETP = cholesterol ester transfer protein; LCAT = lecithin-cholesterol acyltransferase; HL = hepatic lipase; C = cholesterol; CE = cholesterol esters; TG = triglycerides; ABCA1 = ATP-binding cassette A1; SR-B1 = scavenger receptors; CM = chylomicrons; VLDL = very low density lipoprotein; LDL = low density lipoprotein; HDL = high density lipoprotein; ABCG1 = ATP-binding cassette G1

Cholesterol is transported from the arterial wall or other extrahepatic tissues back to the liver by HDL. Esterified cholesterol from HDL can be transferred to apolipoprotein B–containing particles in exchange for triglycerides. Cholesterol esters transferred from HDL to VLDL and LDL are taken up by hepatic LDL receptors or can be delivered back to extrahepatic tissue. (Modified from Ito MK. Hyperlipidemia. In: Chisholm MA, Schwinghammer TL, Wells BG, DiPiro JT, Kolesar JM, Malone PM, eds. Pharmacotherapy: Principles & Practice. New York: McGraw-Hill, 2007:chap. 9; with permission.)

ROLE OF GENETICS

The understanding of genetics in relation to lipid disorders has expanded dramatically in recent years. Many genetic changes (e.g., receptor mutations, genotype variation) can result in specific lipoprotein disorders depending on where in the lipid transport system the interruption occurs. Many factors associated with the increased occurrence of lipoprotein disorders—such as age, sex, gene–gene, and environmental interactions—can make it difficult to directly attribute the sole cause of a patient's lipid abnormalities to inheritable genetic traits.[18] In clinical practice, most lipid disorders encountered are the result of interactions between increased age, decreased or lack of physical activity, weight gain, and suboptimal diet within an individual's genetic makeup.[18] Clinically, genetic lipid disorders can be divided into five main effects, namely changes in LDL-C, Lp(a), remnant lipoproteins, triglyceride-rich lipoproteins (e.g., chylomicrons), and HDL-C.[12] The lipid genetic disorders can either result in an increase or decrease in the amount of lipoprotein present.

FAMILIAL DYSLIPIDEMIAS

Several familial lipid disorders have been associated with an increased risk of coronary artery disease (CAD).[19] Because of the resultant increase in cardiovascular events observed in lipid disorders, there has been a drive to determine the specific classification of these disorders to help coordinate optimal management. Lipid disorders were originally classified by the Frederickson classification system differentiating disorders based on the physical characteristics of lipoproteins (Tables 1 and 2).[20,21] The problem with classifying by Frederickson phenotypes is that this classification system does not permit the diagnosis of lipid disorders because primary and secondary causes are not considered. More recently, lipid disorder classification has focused on identifying disorders based on known or suspected causes (Table 3).

HYPERCHOLESTEROLEMIA
Familial Hypercholesterolemia

Familial hypercholesterolemia (FH) was first recognized in the 1930s when the autosomal codominant means of genetic inheritance was identified.[22] This autosomal dominant disorder is caused by a mutation in the LDL receptor gene. Today, more than 700 mutations in this receptor have been identified.[23] The resultant lack of LDL receptors or dysfunctional LDL receptors interrupts the clearance of LDL and subsequently increases cholesterol concentrations in the plasma. The LDL receptor gene defect can be inherited as either a heterozygous or homozygous gene dysfunction and is found in 1 of every 500 people in the United States. Heterozygotes inherit one defective gene and usually present with LDL-C concentrations of 250–450 mg/dL. Homozygotes inherit a defective LDL receptor gene from both parents, and these patients present with LDL-C concentrations greater than 500 mg/dL. Specific clinical findings in FH are the presence of tendon xanthomas (thickening and cholesterol deposition in extensor tendons), xanthelasmas (cholesterol deposits in the eyelids), arcus cornea (cholesterol infiltration around the corneal rim), and premature CAD.[24]

is higher than single gene mutations, but the specific genetic abnormalities have yet to be defined. In general, the LDL-C elevations are in the range of 130–250 mg/dL. Patients with this disorder generally do not have xanthomas.

HYPERTRIGLYCERIDEMIA

Familial Hypertriglyceridemia

The characteristics of this disorder are an increase in triglycerides enriched with VLDL and chylomicrons. In general, LDL-C is normal, whereas HDL-C is decreased.[22] This disorder is generally mild and asymptomatic unless there are other secondary causes of increased triglycerides. Triglyceride elevations generally occur from 200 to 500 mg/dL. Patients can present with eruptive xanthomas and/or acute pancreatitis with triglyceride concentrations of 500 mg/dL and greater.

Familial LPL Deficiency

A mutation in the LPL gene or a mutation in cofactor apolipoprotein CII is the cause of this rare disorder. Lipemic plasma (milky color secondary to fatty materials in plasma) is observed in most patients and especially in patients with severe disease, when triglyceride concentrations can range from 2000 to 25,000 mg/dL. This disorder usually presents in childhood and can present with pancreatitis, eruptive xanthomas, and lipemia retinalis (creamy-white appearance of retinal blood vessels).[22]

Familial Apolipoprotein CII Deficiency

This disorder is an autosomal recessive inborn error characterized by a deficiency in apolipoprotein CII. The clinical presentation is similar to LPL deficiency because apolipoprotein CII is a cofactor for LPL. Homozygous individuals present with severe hypertriglyceridemia, but heterozygous individuals often have normal plasma lipid values unless multiple genetic abnormalities or lifestyle risk factors are present.[22]

COMBINED HYPERLIPIDEMIA

Familial Combined Hyperlipidemia

Familial combined hyperlipidemia is an autosomal dominant trait that occurs in 1–2% of the population. The general lipid abnormalities seen are a low HDL-C and an increase in LDL-C secondary to an overproduction of apolipoprotein B, resulting in an increase in VLDL production. In general, LDL-C concentrations are 250–350 mg/dL, and triglyceride concentrations are 200–800 mg/dL. The diagnosis is made by identifying a total plasma cholesterol and/or triglyceride concentration greater than the 90th percentile for the general population adjusted for age and sex.[22]

Familial Dysbetalipoproteinemia (Type III)

This disorder is an autosomal recessive trait caused by a mutation of apolipoprotein E that results in a dysfunction with the binding of lipoprotein receptors. Apolipoprotein E is required in the normal clearance of VLDL and chylomicrons. Three isoforms of apolipoprotein E exist: E2, E3, and E4. Of the three isoforms mentioned, the E2 iso-

TABLE 2. FREDERICKSON CLASSIFICATION OF DYSLIPIDEMIAS[20,21]

Phenotype	Elevated Lipoprotein(s)	Plasma TC	Plasma TG	Estimated Prevalence (%)
I	Chylomicrons	Normal or ↑	↑↑↑	< 1.0
IIa	LDL	↑↑	Normal or ↑	20–80
IIb	LDL and VLDL	↑↑	↑↑	10
III	IDL	↑↑	↑	0.02
IV	VLDL	↑	↑↑	1
V	VLDL and chylomicrons ↑		↑↑	0.1

IDL = intermediate-density lipoprotein; LDL = low-density lipoprotein; TC = total cholesterol; TG = triglycerides; VLDL = very low-density lipoprotein; ↑ = indicates increased concentrations.

TABLE 3. CLASSIFICATION OF FAMILIAL DYSLIPIDEMIA[22]

↑ Cholesterol	↑ TG	↑ Cholesterol and TG	↓ HDL-C	↑ HDL-C
Familial hypercholesterolemia	Familial hypertriglyceridemia	Familial combined hyperlipidemia	Familial hypoalphalipoproteinemia	Familial hyperalphalipoproteinemia
Polygenic hypercholesterolemia	Lipoprotein lipase deficiency	Familial dysbetalipoproteinemia	Apolipoprotein A1 deficiency	CETP deficiency
Familial defective apolipoprotein B-100	Apolipoprotein CII deficiency		LCAT deficiency	Apolipoprotein A1 overexpression
Familial combined hyperlipidemia			Fish-eye disease Tangier disease	

Source: Hachem SB, Mooradian AD. Familial dyslipidaemias: an overview of genetics, pathophysiology and management. Drugs 2006;66:1949–69. Permission to use obtained from Wolters Kluwer Health Publications.
CETP = cholesterol ester transport protein; HDL-C = high-density lipoprotein cholesterol; LCAT = lecithin cholesterol acyltransferase; TG = triglycerides; ↑ = increase; ↓ = decrease.

Familial Defective Apolipoprotein B-100

A mutation in the apolipoprotein B-100 is the cause of familial defective apolipoprotein B-100 and results in an inability of LDL to bind to the receptor. Patients with familial defective apolipoprotein B-100 present with an increase in LDL-C concentrations (usually 250–450 mg/dL) secondary to an inability to clear LDL-C.[22] The clinical presentation is indistinguishable from heterozygous FH and can present with the following manifestations of tendon xanthomas, xanthelasmas, and premature CAD.

Polygenic Hypercholesterolemia

This disorder is a combination of genetic metabolism abnormalities and environmental influence (e.g., poor dietary habits, sedentary lifestyle) contributing to the resultant increase in LDL-C. Polygenic hypercholesterolemia is the most common cause of increased LDL-C and appears in about 25% of the U.S. population. The prevalence form has the lowest binding affinity to the LDL receptor and results in the most lipid abnormalities in individuals with the apoE2:E2 phenotype. Low-density lipoprotein cholesterol concentrations generally range from 300 to 600 mg/dL, whereas triglyceride concentrations range from 400 to 800 mg/dL. The primary clinical presentation consists of xanthomas, but alternative presentations can include CAD and peripheral artery disease. The presence of planar palmar xanthomas provides a high diagnostic predictive value for familial dysbetalipoproteinemia.[22] The general lipid profile is characterized by elevated triglycerides, normal HDL-C, and low LDL-C.

Familial HL Deficiency

Hepatic lipase is responsible for removing triglycerides and phospholipids from chylomicron and VLDL remnants.[22] Patients with HL deficiency present with severe hypertriglyceridemia, modestly elevated LDL-C, and normal to increased HDL-C.

DISORDERS OF HDL METABOLISM

Familial Hypoalphalipoproteinemia

This disorder is an autosomal dominant disorder in which the specific genetic defect has yet to be determined. The main manifestation of this disorder is low HDL-C (less than 30 mg/dL in men, less than 40 mg/dL in women) in the plasma. Specific ethnic groups, especially individuals of American Indian ancestry, are the main population with this disorder.

Apoprotein AI Deficiency

Apoprotein AI deficiency is characterized by a very low HDL-C (less than 10 mg/dL) secondary to a mutation in the apolipoprotein AI gene. Clinical findings include xanthomas, corneal opacities, and premature CAD.[22]

LCAT Deficiency

Lecithin cholesterol acyltransferase deficiency is a rare recessive disorder. The main deficiency is a decreased cholesterol esterification to cholesteryl ester on HDL particles. This disorder leads to free cholesterol accumulation in peripheral tissues that can result in corneal opacities, normochromic anemia, and kidney failure.[22] Fish-eye disease is a variant of LCAT deficiency resulting from an LCAT gene mutation, but it generally causes less severe clinical manifestations.

Tangier Disease

Tangier disease is an autosomal recessive disorder that causes increased catabolism of HDL-C in the plasma, resulting in a lipid profile characterized by low LDL-C, low HDL-C, and increased triglycerides. The ABC transport 1-gene mutations have been linked to this disease because ABCA1 is responsible for the normal exchange of cholesterol from peripheral tissues. Clinical manifestations typically include orange tonsils, corneal deposits, hepatomegaly, splenomegaly, peripheral neuropathy, and premature CAD.[23]

Familial Hyperalphalipoproteinemia

This disorder is characterized by elevations in HDL-C usually greater than the 90th percentile based on age and sex.[25] It occurs either through a deficiency in CETP or through idiopathic causes. A deficiency in CETP generally also results in an increased plasma HDL-C and a decreased risk of CAD.

DISORDER OF LP(A) EXCESS

Similar to LDL, each Lp(a) particle has one molecule of apolipoprotein B-100. In addition, each Lp(a) particle contains an apolipoprotein (a) linked to apolipoprotein B-100 by a single disulfide bond. The atherogenicity of Lp(a) is unclear secondary to the varying apolipoprotein (a) isoforms and the unknown differences in atherogenic risk among individual isoforms. The various sizes of apolipoprotein (a) also make accurate measurements of serum Lp(a) a challenge. Apolipoprotein (a) is structurally similar to plasminogen, but it has no fibrinolytic properties and likely interferes with the physiologic role of plasminogen. The atherogenicity of Lp(a) might reside in its preferential binding to fibrin, leading to increased lipid retention and accumulation in the arterial wall.[26] Increases (greater than 30 mg/dL) in Lp(a) are associated with increases in the risk of CAD.[27] However, general screening for Lp(a) is currently not recommended.

SECONDARY LIPID DISORDERS

Secondary causes of lipid disorders often accompany genetic lipid disorders or predispose individuals to abnormal lipid concentrations. An easy way to remember the secondary causes of lipid disorders is by the 4D's mnemonic of diet, drugs, disorders, and diseases.[28] The commonly reported secondary causes of dyslipidemia and their subsequent lipoprotein change are listed in Table 4.

DIET

Low-density lipoproteins can be elevated by the ingestion of saturated fats, excessive calorie intake, and overall high dietary cholesterol consumption. In addition, dietary factors can result in the lowering of cholesterol concentrations in patients. Examples of these dietary factors include implementing a daily intake of soluble fiber in the diet and substituting unsaturated fats or complex carbohydrates for saturated fats.[29] Consumption of alcohol in the diet can result in an elevation of triglycerides as well as HDL-C. The Adult Treatment Panel III (ATP III) guidelines and 2004 update have specific recommendations related to dietary consumption habits in patients with lipid abnormalities.[16,29,30] A key patient population to be aware of is patients with anorexia nervosa who have extreme weight losses, which often result in an alarming hypercholesterolemia.[31,32]

DRUGS

Diuretics and β-blockers have been linked to lipid abnormalities—primarily increases in total cholesterol and increases in triglycerides with subsequent decreases in HDL-C.[16,33] Estrogens and glucocorticoids are also major contributors to secondary causes

TABLE 4. 4D'S CLASSIFICATION OF THE SECONDARY CAUSES OF DYSLIPIDEMIA[16,22,28-43]

Cause	Lipoprotein Changes
Diet	
Alcohol	↑ TG, ↑ HDL-C
Anorexia nervosa	↑ LDL-C
Weight gain	↑ TG, ↓ HDL-C
Drugs	
Amiodarone	↑ LDL-C
Anabolic steroids	↑ LDL-C, ↓ HDL-C
α-Blockers	↑ HDL-C
Anti-epileptics	↑ HDL-C
Atypical antipsychotics	↑ TG, ↑ or NC in LDL-C
β-Blockers	↑ TG, ↓ HDL-C
Conventional antipsychotics	↑ TG, ↑ or NC in LDL-C
Cyclosporine	↑ LDL-C, ↑ TG
Exogenous estrogen	↑ TG, ↓ LDL-C, ↑ HDL-C
Glucocorticoids	↑ to NC in LDL-C, ↑ or NC in HDL-C, ↑ HDL-C
Interferon	↑ TG
Pioglitazone	↓ TG, ↑ LDL-C, ↑ HDL-C
Protease inhibitors	↑ LDL-C, ↑ TG, ↓ HDL-C
Retinoid acid derivatives	↑ LDL-C, ↑ TG, ↓ HDL-C
Rosiglitazone	↑ TG, ↑ LDL-C, ↑ HDL-C
Thiazide diuretics	↑ LDL-C, ↑ or NC in TG
Disorder (metabolic)	
Acromegaly	↑ TG
Diabetes mellitus	↑ TG, ↓ HDL-C
Glycogen storage disorders	↑ TG
Hypothyroidism	↑ LDL-C
Lipodystrophy	↑ TG, ↓ HDL-C
Polycystic ovary syndrome	↑ TG, ↑ LDL-C, ↓ HDL-C
Pregnancy	↑ LDL-C, ↑ TG
Disease (nonmetabolic)	
Burns	↑ TG
Chronic kidney disease	↑ LDL-C, mixed hyperlipidemia
Myelomatosis	↑ LDL-C
Nephrotic syndrome	↑ LDL-C, ↓ HDL-C
Obstructive liver disease	↑ LDL-C, ↑ lipoprotein X
Porphyria	↑ LDL-C
Systemic lupus erythematosus	↑ TG

HDL-C = high-density lipoprotein cholesterol; LDL-C = low-density lipoprotein cholesterol; NC = no change; TG = triglycerides; ↑ = indicates increased concentrations; ↓ = indicates decreased concentrations.

of dyslipidemia and primarily result in increases in triglycerides through increases in VLDL production and HDL-C.[34] Low-potency conventional antipsychotics (e.g., chlorpromazine, thioridazine) and atypical antipsychotics (e.g., clozapine, olanzapine, quetiapine) are associated with a high risk of hyperlipidemia, predominantly hypertriglyceridemia.[35] Certain drugs used to treat diabetes can also affect lipid concentrations in patients, specifically insulin (decreased LDL-C) and thiazolidinediones such as rosiglitazone (increased triglycerides, HDL-C, and LDL-C) and pioglitazone (decreased triglycerides, increased HDL-C, and increased LDL-C).[36,37] Other drugs that may cause lipid abnormalities are amiodarone, cyclosporine, anticonvulsants, protease inhibitors, and retinoids.[28]

DISORDERS OF METABOLISM

Changes in or dysfunction of the regular metabolic pathways can result in lipid abnormalities. Hypothyroidism is one of the most common secondary disorders that can cause hypercholesterolemia, specifically increases in both total cholesterol and LDL-C.[38] Dramatic changes in the lipid profiles of obese patients, especially increases in triglycerides and decreases in HDL-C, result in a multitude of cardiovascular consequences. Type 2 diabetes mellitus has been associated with hypertriglyceridemia, low HDL-C, and other abnormalities in lipoprotein particle size and physical properties.[39] Polycystic ovary syndrome (PCOS) generally presents with elevated triglycerides and LDL-C and decreased HDL-C.[40] Pregnancy is the last of the major disorders that can affect the metabolism of cholesterol. Plasma cholesterol concentrations can increase by 50% during the second trimester of pregnancy,[28] whereas triglycerides can increase by up to three times their normal concentration during pregnancy, usually peaking during the third trimester because of the overproduction of triglyceride-rich VLDL.[41]

DISEASES

Liver disease resulting in hypercholesterolemia can be related to cholestasis or primary biliary cirrhosis. Any hepatocellular damage can result in an impairment in the hepatic production of the apolipoproteins and enzymes involved in lipid metabolism.[28] Patients with chronic kidney disease (CKD) according to the Kidney Disease Outcome Quality Initiative (K/DOQI) guidelines (defined by a glomerular filtration rate [GFR] less than 60 mL/minute/1.73 m^2 for 3 months or longer) have increased triglycerides and low HDL-C.[42] The makeup of triglycerides in chronic renal failure often represents atherogenic lipoproteins, such as chylomicrons, VLDL remnants, IDLs, and small, dense LDLs.[43] One of the main characteristics of nephrotic syndrome is hypercholesterolemia (e.g., increased total cholesterol, increased LDL-C, increased VLDL), but hypertriglyceridemia can also be present later in nephrotic syndrome when albumin concentrations are decreased.[43] Following solid organ transplantation patients often have multiple factors contributing to hyperlipidemias, which include some of the previously mentioned secondary causes such as weight gain, steroids, cyclosporine, antihypertensive use, and the development of metabolic syndrome.[28]

REFERENCES

1. Libby P. Molecular basis of the acute coronary syndrome. Circulation 1995;91:2844–50.
2. Shah PK, Falk E, Badimon JJ, et al. Human monocyte-derived macrophages induce collagen breakdown in fibrous caps of atherosclerotic plaques: potential role of matrix-degrading metalloproteinases and implications for plaque rupture. Circulation 1995;92:1565–9.
3. Khalil MF, Wagner WD, Goldberg IJ. Molecular interactions leading to lipoprotein retention and the initiation of atherosclerosis. Arterioscler Thromb Vasc Biol 2004;24:2211–8.
4. Leitinger N. Oxidized phospholipids as modulators of inflammation in atherosclerosis. Curr Opin Lipidol 2003;14:421–30.
5. Agirbasli M. Pivotal role of plasminogen-activator inhibitor 1 in vascular disease. Int J Clin Pract 2005;59:102–6.
6. Tesfamariam B, Defelice AF. Endothelial injury in the initiation and progression of vascular disorders. Vascul Pharmacol 2007;46:229–37.
7. Tobias P, Curtiss LK. Thematic review series: the immune system and atherogenesis. Paying the price for pathogen protection: toll receptors in atherogenesis. J Lipid Res 2005;46:404–11.
8. Rader DJ, Pure E. Lipoproteins, macrophage function, and atherosclerosis: beyond the foam cell? Cell Metab 2005;1:223–30.
9. de Winther MP, Kanters E, Kraal G, Hofker MH. Nuclear factor kappaB signaling in atherogenesis. Arterioscler Thromb Vasc Biol 2005;25:904–14.
10. Tabas I. Consequences and therapeutic implications of macrophage apoptosis in atherosclerosis: the importance of lesion stage and phagocytic efficiency. Arterioscler Thromb Vasc Biol 2005;25:2255–64.
11. Brown BG, Zhao XQ, Chait A, et al. Simvastatin and niacin, antioxidant vitamins, or the combination for the prevention of coronary disease. N Engl J Med 2001;345:1583–92.
12. Genest J. Lipoprotein disorders and cardiovascular risk. J Inherit Metab Dis 2003;26:267–87.
13. Cella LK, Van Cauter E, Schoeller DA. Effect of meal timing on diurnal rhythm of human cholesterol synthesis. Am J Physiol 1995;269(5 pt 1):E878–83.
14. Turley SD, Dietschy JM. The intestinal absorption of biliary and dietary cholesterol as a drug target for lowering the plasma cholesterol level. Prev Cardiol 2003;6:29–33, 64.
15. Ito MK, Talbert RL, Tsimikas S. Statin-associated pleiotropy: possible beneficial effects beyond cholesterol reduction. Pharmacotherapy 2006;26(7 pt 2):85S–97S.
16. Third Report of the National Cholesterol Education Program (NCEP) Expert Panel on Detection, Evaluation, and Treatment of High Blood Cholesterol in Adults (Adult Treatment Panel III) final report. Circulation 2002;106:3143–421.
17. Garcia-Calvo M, Lisnock J, Bull HG, et al. The target of ezetimibe is Niemann-Pick C1-Like 1 (NPC1L1). Proc Natl Acad Sci USA 2005;102:8132–7.

18. Genest J, Libby P, Gotto AM. Lipoprotein disorders and cardiovascular disease. In: Zipes DP, Libby P, Bonow RO, Braunwald E, eds. Braunwald's Heart Disease: A Textbook of Cardiovascular Medicine, 7th ed. Philadelphia: Elsevier Saunders, 2005:1013-32.
19. Genest JJ, Martin-Munley SS, McNamara JR, et al. Familial lipoprotein disorders in patients with premature coronary artery disease. Circulation 1992;85:2025-33.
20. Frederickson DS, Levy RI, Lees RS. Fat transport in lipoproteins—an integrated approach to mechanisms and disorders. N Engl J Med 1967;276:148-56.
21. Hirschhorn K, Wilkinson CR. The mode of inheritance in essential familial hypercholesterolemia. Am J Med 1959;26:60.
22. Hachem SB, Mooradian AD. Familial dyslipidaemias: an overview of genetics, pathophysiology and management. Drugs 2006;66:1949-69.
23. Serfaty-Lacrosniere C, Civeira F, Lanzberg A, et al. Homozygous Tangier disease and cardiovascular disease. Atherosclerosis 1994;107:85-98.
24. Yuan F, Wang J, Hegele RA. Heterozygous familial hypercholesterolemia: an underrecognized cause of early cardiovascular disease. CMAJ 2006;174:1124-9.
25. Yamashita S, Maruyama T. Molecular mechanisms, lipoprotein abnormalities and atherogenicity of hyperalphalipoproteinemia. Atherosclerosis 2000;152:271-85.
26. Smith EB, Crosbie L. Does lipoprotein (a) [Lp(a)] compete with plasminogen in human atherosclerotic lesions and thrombi? Atherosclerosis 1991;89:127-36.
27. Anuurad E, Boffa MB, Koschinsky ML, Berglund L. Lipoprotein (a): a unique risk factor for cardiovascular disease. Clin Lab Med 2006 26:751-72.
28. Stone NJ. Secondary causes of hyperlipidemia. Med Clin North Am 1994;78:117-41.
29. Van Aalst-Cohen ES, Jansen AC. Clinical, diagnostic, and therapeutic aspects of familial hypercholesterolemia. Semin Vasc Med 2004;4:31-41.
30. Grundy SM, Cleeman JI, Merz CN, et al. Implications of recent clinical trials for the National Cholesterol Education Program Adult Treatment Panel III guidelines. Circulation 2004;110:227-39.
31. Klinefelter HF. Hypercholesterolemia, in anorexia nervosa. J Clin Endocrinol Metab 1965;25:1520-3.
32. Phinney SD, Tang AB, Waggoner CR, et al. The transient hypercholesterolemia of major weight loss. Am J Clin Nutr 1991;53:1404-10.
33. Henkin Y, Como JA, Oberman A. Secondary dyslipidemia: inadvertent effects of drugs in clinical practice. JAMA 1992;267:961-8.
34. The Writing Group for the PEPI Trial. Effects of estrogens or estrogen/progestin regimens on heart disease risk factors in postmenopausal women. The Postmenopausal Estrogen/Progestin Interventions (PEPI) Trial. JAMA 1995;273:199-208.
35. Meyer JM, Koro CE. The effects of antipsychotic therapy on serum lipids: a comprehensive review. Schizophr Res 2004;70:1-17.
36. Fonseca V, Rosenstock J, Patwardhan R, Salzman A. Effect of metformin and

rosiglitazone combination therapy in patients with type 2 diabetes mellitus: a randomized controlled trial. JAMA 2000;283:1695-702.
37. Goldberg RB, Kendall DM, Deeg MA, et al. A comparison of lipid and glycemic effects of pioglitazone and rosiglitazone in patients with type 2 diabetes and dyslipidemia. Diabetes Care 2005;28:1547-54.
38. Ball MJ, Griffiths D, Thorogood M. Asymptomatic hypothyroidism and hypercholesterolemia. J R Soc Med 1991;84:527-9.
39. Abate N, Vega GL, Garg A, Grundy SM. Abnormal cholesterol distribution among lipoprotein fractions in normolipidemic patients with mild NIDDM. Atherosclerosis 1995;118:111-22.
40. Pirwany IR, Fleming R, Greer IA, et al. Lipids and lipoprotein subfractions in women with PCOS: relationship to metabolic and endocrine parameters. Clin Endocrinol 2001;54:447-53.
41. Humphrey JL, Childs MT, Montes A, Knopp RH. Lipid metabolism in pregnancy. VII. kinetics of chylomicron triglyceride removal in fed pregnant rat. Am J Physiol 1980;239:E81-E87.
42. National Kidney Foundation. K/DOQI clinical practice guidelines for chronic kidney disease: evaluation, classification, and stratification. Kidney Disease Outcome Quality Initiative. Am J Kidney Dis 2002;39(suppl 2):S1-S266.
43. Grundy SM. Management of hyperlipidemia of kidney disease. Kidney Int 1990;37:847-53.

2
CLINICAL EVALUATION OF THE ADULT PATIENT

Kim K. Birtcher, M.S., Pharm.D., BCPS (AQ Cardiology), CDE, CLS

IDENTIFY DYSLIPIDEMIAS[1,2]

Abnormalities in individual lipid parameters have been associated with coronary heart disease (CHD). Elevated concentrations of LDL-C and low concentrations of HDL-C have been associated with an increased risk of cardiovascular events. Elevated triglyceride concentrations have been associated with small, dense LDL-C particles, which also increase the risk of cardiovascular events.

DETERMINE LIPID PROFILES

Measuring the fasting lipid profile is an important initial step in determining a patient-specific treatment plan. The information provided in a lipid profile should be used in conjunction with the patient's overall risk of cardiovascular disease (CVD) to determine the treatment strategy.

Assessment for patients with no clinically apparent atherosclerotic disease:

Beginning at age 20, perform a lipoprotein profile after a 9- to 12-hour fast.
- If normal, repeat at least once every 5 years.
- If abnormal, screen for secondary causes of dyslipidemia (suggested laboratory tests: fasting blood glucose, thyroid-stimulating hormone, liver function tests, serum creatinine [SCr], and urinalysis).
  - Hypothyroidism

TABLE 1. CLASSIFICATION OF SERUM LIPOPROTEINS

Total cholesterol (mg/dL)
Desirable	< 200
Borderline-high	200–239
High	> 240

LDL-C (mg/dL)
Optimal	< 100
Above/near optimal	100–129
Borderline high	130–159
High	160–189
Very high	> 190

HDL-C (mg/dL)
Low	
Men	< 40
Women	< 50[3]
High	> 60

Triglycerides (mg/dL)
Normal	< 150 mg/dL
Borderline high	150–199 mg/dL
High	200–500 mg/dL
Very high	> 500 mg/dL

- - Obstructive liver disease
  - CKD
  - Drugs that influence lipid panel (e.g., progestins, estrogens, anabolic steroids, corticosteroids, protease inhibitors, isotretinoin)
  - Diet, alcohol abuse
- If a nonfasting lipoprotein profile is performed, only the total cholesterol and HDL-C are accurate. In the nonfasting state, the triglycerides will be higher and the LDL-C will be lower than in the fasting state. Because treatment guidelines are based on LDL-C values, it is best to obtain a fasting lipid panel.
  - If total cholesterol is 200 mg/dL or HDL-C is less than 40 mg/dL, obtain a fasting lipoprotein profile.
- With severe hypercholesterolemia (e.g., total cholesterol greater than 300 mg/dL; LDL-C greater than 190 mg/dL), obtain family history and screen family members for hypercholesterolemia.
- Use Table 1 to classify the components of the lipid panel. This classification system is useful when writing the progress note for the patient. Base the treatment intensity on the patient's overall risk, not on this classification system.

RISK STRATIFICATION

The intensity of lipid-lowering therapy is determined by the patient's risk of a vascular event. Patients with the highest risk of a cardiovascular event will require the most aggressive therapy. When the patient has two or more risk factors for CHD and does not have clinical CHD or CHD risk equivalent, the modified Framingham Risk Score should be used to determine the patient's 10-year risk of a CHD event and the initial treatment strategy. The Framingham Risk Score should not be used to determine a patient's 10-year risk of a CHD event if the patient has clinical CHD, CHD risk equivalent, or one or no National Cholesterol Education Program (NCEP) risk factor. The risk stratification steps are summarized in Figure 1.

Step 1: Does the patient have clinical CHD or is the patient a CHD risk equivalent?
Clinical CHD:
- Prior myocardial infarction
- Prior revascularization (bypass surgery/percutaneous coronary intervention)
- Silent ischemia or angina pectoris

CHD risk equivalent:
- Other vascular disease caused by atherosclerosis
  - Peripheral artery disease
  - Carotid artery disease (history of an ischemic stroke or transient ischemic attack)[3]
  - Abdominal aortic aneurysm
  - Renal artery stenosis
- Diabetes
- 20% 10-year risk of a CHD event (from Framingham Risk Scores)
- CKD[4]

If YES, go to Step 4.

FIGURE 1. RISK STRATIFICATION

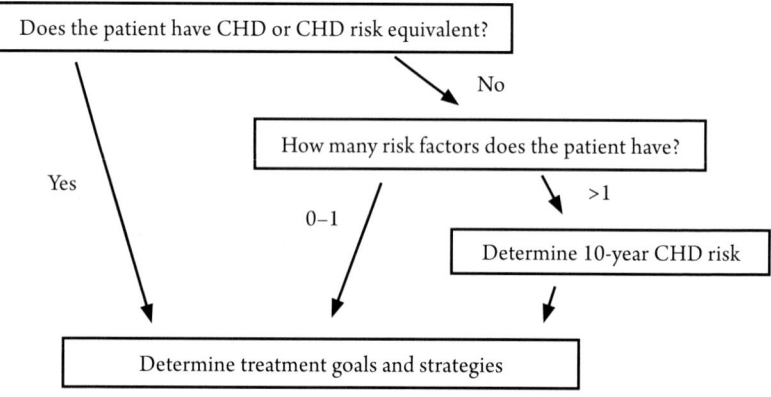

Step 2: How many major risk factors does the patient have?
- Age (men age 45 and older; women age 55 and older)
- Current cigarette smoking
- Hypertension (blood pressure 140/90 mm Hg or higher or on antihypertension drug)
- HDL (men less than 40 mg/dL; women less than 50 mg/dL.[5] If HDL is 60 mg/dL or higher, this is desirable and counts as a negative risk factor; subtract 1 point)
- Family history of premature CHD (men, first-degree relative younger than 55; women, first-degree relative younger than 65)

If the patient has one risk factor or none, go to Step 4.
If the patient has two or more risk factors, go to Step 3.

Step 3: Use Table 2 to determine the 10-year CHD risk when the patient has two or more risk factors and does not have clinical CHD or CHD risk equivalent status.

Electronic calculators are available at www.nhlbi.nih.gov/guidelines/cholesterol. The electronic calculator may provide slightly different point totals. The Framingham Risk Score may underestimate the risk of cardiovascular events in women;[5] the average lifetime risk of CVD in women is one in two.

Step 4: Determine treatment goals and strategies.
- Establish the LDL-C goal (Table 3).
- Determine whether to start therapeutic lifestyle changes (TLCs).
- Determine whether to start drug therapy.

Lowering the LDL-C is the primary goal of therapy, unless the triglycerides are more than 500 mg/dL. (When the triglycerides are more than 500 mg/dL, give priority to lowering the triglycerides to reduce the patient's risk of pancreatitis.) When the LDL-C goal has been achieved, focus first on achieving the non-HDL goal and then the HDL-C target value.
- The non-HDL goal is 30 mg/dL higher than the LDL-C goal (i.e., when the LDL-C goal is 100 mg/dL, the non-HDL goal is 130 mg/dL).
- The patient's non-HDL is determined by subtracting the HDL from the total cholesterol (non-HDL = total cholesterol − HDL).

Advanced Lipid Tests

There is no specified role for advanced lipid testing in the current NCEP guidelines. There may be value in using these tests in patients with any of the following:
- Unremarkable lipid panel results, two or more NCEP risk factors, and a strong family history of CHD;
- A personal history of a premature CHD event with no apparent explanation; or
- A history of recurrent CVD despite optimal management with TLCs and drug therapy.

The additional information that is provided by the advanced lipid tests (Table 4) might help justify more aggressive, targeted lipid management for patients with bor-

TABLE 2. MODIFIED FRAMINGHAM RISK SCORES FOR MEN AND WOMEN

Men				Women			
Age	Points	Age	Points	Age	Points	Age	Points
20–34	–9	55–59	8	20–34	–7	55–59	8
35–39	–4	60–64	10	35–39	–3	60–64	10
40–44	0	65–69	11	40–44	0	65–69	12
45–49	3	70–74	12	45–49	3	70–74	14
50–54	6	75–79	13	50–54	6	75–79	16

TC (mg/dL)	Age (years)					Age (years)				
	20–39	40–49	50–59	60–69	70–79	20–39	40–49	50–59	60–69	70–79
< 160	0	0	0	0	0	0	0	0	0	0
160–199	4	3	2	1	0	4	3	2	1	1
200–239	7	5	3	1	0	8	6	4	2	1
240–279	9	6	4	2	1	11	8	5	3	2
≥ 280	11	8	5	3	1	13	10	7	4	2

Age	20–39	40–49	50–59	60–69	70–79	20–39	40–49	50–59	60–69	70–79
Nonsmoker	0	0	0	0	0	0	0	0	0	0
Smoker	8	5	3	1	1	9	7	4	2	1

HDL-C (mg/dL)	Points	HDL-C (mg/dL)	Points
≥ 60	–1	≥ 60	–1
50–59	0	50–59	0
40–49	1	40–49	1
< 40	2	< 40	2

Systolic BP (mm Hg)	Untreated	Treated	Systolic BP (mm Hg)	Untreated	Treated
< 120	0	0	< 120	0	0
120–129	0	1	120–129	1	3
130–139	1	2	130–139	2	4
140–159	1	2	140–159	3	5
≥ 160	2	3	≥ 160	4	6

Point total	10-year risk (%)	Point total	10-year risk (%)
0	1	< 9	< 1
1	1	9	1
2	1	10	1
3	1	11	1
4	1	12	1
5	2	13	2
6	2	14	2
7	3	15	3
8	4	16	4
9	5	17	5
10	6	18	6
11	8	19	8
12	10	20	11
13	12	21	14
14	16	22	17
15	20	23	22
16	25	24	27
≥ 17	≥ 30	≥ 25	≥ 30

BP = blood pressure; TC = total cholesterol.

TABLE 3. LDL-C GOALS AND CUT POINTS FOR THERAPEUTIC LIFESTYLE CHANGES (TLCS) AND DRUG THERAPY BASED ON RISK CATEGORY

Risk Category	LDL-C Goal (mg/dL)	Start TLCs[a] (mg/dL)	Start Drug Therapy[b] (mg/dL)
High risk CHD or CHD risk equivalent; 10-year risk > 20%	< 100 (optional < 70)[d]	> 100[c]	> 100 (< 100 – consider drug)[e]
Moderately high risk 2+ risk factors; 10-year risk 10–20%	< 130 (optional < 100)[f]	> 130[c]	> 130 (100–129 – consider drug)[g]
Moderate risk 2+ risk factors; 10-year risk < 10%	< 130	> 130	> 160
Lower risk 0 or 1 risk factor	< 160	> 160	> 190

[a]Dietary modification, weight reduction, exercise.
[b]When using LDL-C–lowering therapy, achieve at least a 30–40% LDL-C reduction.
[c]Regardless of LDL-C, lifestyle changes are indicated when lifestyle-related risk factors (obesity, physical inactivity, ↑ triglycerides, ↓ HDL, or metabolic syndrome) are present.
[d]Consider an LDL-C goal of less than 70 mg/dL for any patient with events or symptomatic atherosclerotic disease.[6]
[e]If baseline LDL-C is less than 100 mg/dL, LDL-C–lowering therapy is an option based on clinical trials. With ↑ triglycerides or ↓ HDL, consider combining fibrate or nicotinic acid with an LDL-C–lowering drug.
[f]Consider LDL-C goal less than 100 mg/dL for patients of advanced age; three or more risk factors; current tobacco use; triglycerides 200 mg/dL or greater and non–HDL-C of 160 mg/dL or more; low HDL-C; metabolic syndrome; and/or emerging risk markers (e.g., high-sensitivity C-reactive protein [hs-CRP], Lp(a), coronary calcium greater than 75th percentile for age/sex).
[g]At baseline or after lifestyle changes—initiating therapy to achieve LDL-C less than 100 mg/dL is an option based on clinical trials.
CHD = coronary heart disease. Coronary heart disease includes a history of myocardial infarction, unstable angina, stable angina, coronary artery revascularization procedures (angioplasty or bypass surgery), and evidence of clinically significant myocardial ischemia. Coronary heart disease risk equivalent includes peripheral artery disease, carotid artery disease (history of ischemic stroke or transient ischemic attack), abdominal aortic aneurysm, renal artery stenosis, diabetes, chronic kidney disease,[5] or more than a 20% 10-year risk of a CHD event based on the modified Framingham Risk Score.
Adapted from the National Cholesterol Education Program[1] and NCEP Report.[2] All 10-year risks are based on Framingham stratification (electronic calculators are available at www.nhlbi.nih.gov/guidelines/cholesterol).

derline lipid values. For example, knowing that a patient has an elevated Lp(a); small, dense LDL-Cs; VLDL3; and/or large numbers of LDL particles justifies being more aggressive with lipid treatment. This information may be used to motivate the patient to comply with TLCs and medication therapy. The clinician can use pre- and postintervention advanced lipid testing to motivate the patient to continue TLCs and/or drug therapy that has been prescribed.

The VAP Cholesterol Test[7]

The VAP (vertical auto profile) cholesterol test provides additional information not available with the standard lipid profile. The VAP test directly measures HDL-C, LDL-C, triglycerides, Lp(a), IDL-C, LDL-C size and subtypes, HDL-C subtypes,

TABLE 4. PARAMETERS PROVIDED BY THE LIPID PANEL, VAP TEST, AND LIPOPROFILE TEST

Lipid Panel	VAP Test	LipoProfile Test
TC	TC	TC
LDL-C (calculated using the Friedewald equation)	LDL-C (directly measured)	LDL-C (directly measured)
HDL-C	HDL-C	HDL-C
TG	TG	TG
	Non–HDL-C (LDL-C+VLDL-C)	VLDL subclass particle number
	Lp(a)	LDL subclass particle number
	Remnant lipoproteins (IDL+VLDL3)	HDL subclass particle number
	LDL-C density pattern	
	HDL-2	
	HDL-3	
	VLDL-C, total	
	VLDL-3	
	IDL-C	
	apoB$_{100}$	

TC = total cholesterol.

and VLDL-C-3. It calculates total cholesterol, non-HDL-C (LDL-C + VLDL-C), and total apoB$_{100}$. It also provides some interpretation of the lipid subtypes and graphically shows the LDL-C density pattern.

LipoProfile Test[8]

The *LipoProfile* test uses nuclear magnetic resonance (NMR) spectroscopy to measure the LDL, HDL, and VLDL particle numbers directly; LDL particle size; and standard lipid profile components. It also provides some interpretation of the results and graphically presents the subclass particle numbers by size of particle. Higher numbers of LDL particles are associated with greater CHD risk. The number of LDL particles cannot be reliably predicted using LDL-C alone. Patients may have increased risk of CHD because of high numbers of LDL particles but may have a low or normal LDL-C concentration.

EMERGING RISK MARKERS[1,9,10]

Several emerging risk markers may be useful in identifying patients at high risk of CVD. With the current NCEP guidelines, emerging risk markers are not routinely measured to determine treatment goals. Their measurement may be helpful in determining the need for and intensity of therapy in patients with two or more NCEP risk factors whose lipid panel is otherwise unremarkable. The more promising emerging risk markers are discussed in Table 5.

TABLE 5. EMERGING RISK MARKERS

Emerging Risk Marker and Key Points	Increased By	Decreased By	Measurement and Interpretation[9]
High-sensitivity C-reactive protein (hs-CRP) • Is a mediator of immune response • Serves as a marker of inflammation • May be stronger CHD risk predictor than LDL-C • Do not know if there are clinical benefits from lowering hs-CRP • May be useful in motivating a patient to comply with lifestyle modifications (i.e., patient with unremarkable lipid panel needing lifestyle modification) • Use selectively (i.e., ↑LDL-C, > 2 risk factors, and 10-year risk 10–20%; if hs-CRP ≥ 1 mg/L, use more intensive treatment) • A value > 2 mg/dL is a strong predictor of CHD risk	Inflammation Smoking Hormone replacement therapy	Weight loss Physical activity Statins Fibrates Ezetimibe Aspirin	• Obtain two hs-CRP measurements • Use lower or average • Patient must be clinically stable (no infection, recent hospitalization, recent trauma) • If > 10 mg/L, repeat test; probably noncardiac cause of elevation. Chronic inflammatory diseases (e.g., rheumatoid arthritis) will increase hs-CRP concentrations Risk level of hs-CRP • Low: < 1.0 mg/L • Intermediate: 1.0–3.0 mg/L • High: > 3 mg/L
Fibrinogen • Many factors modulate fibrinogen concentrations (e.g., diabetes, hypertension, inflammation, obesity, sedentary lifestyle, smoking) • Elevated concentrations associated with cardiovascular events • Unknown whether it has causal role or is a marker of vascular damage • Unknown whether there are clinical benefits to lowering fibrinogen	Inflammation Smoking	Exercise Stop smoking Niacin Fibrates	No recommendations at this time

(Continued)

TABLE 5. EMERGING RISK MARKERS (*Continued*)

Emerging Risk Marker and Key Points	Increased By	Decreased By	Measurement and Interpretation[9]
Homocysteine • Patients with inborn errors in homocysteine metabolism have a high risk of venous thrombosis at an early age (< 30 years). Risk may be partially decreased with high-dose B vitamins • Mild-moderate elevations may predispose patients to atherosclerosis. • Not known whether it has causal role or is a marker of vascular damage • Not known if there are cardiovascular benefits to lowering homocysteine. Lowering homocysteine with folic acid may reduce the risk of stroke[11] but has not been consistently associated with other cardiovascular risk reduction[12]	Renal failure Hypothyroidism Deficiency of folate, B_6, B_{12} Methotrexate Genetics	Replacement of folate, B_6, B_{12} Genetics	No recommendations at this time
Lp(a) • LDL-like particle controlled mostly by genetics • May be highly thrombogenic • When elevated, may increase the risk of cardiovascular events in patients with diabetes, hypertension, ↑ LDL-C, ↓ HDL-C, ↑ homocysteine, ↑ fibrinogen • Unaffected by diet or exercise • Do not know if there are clinical benefits to lowering Lp(a) • Optimal role in screening/risk determination has not been determined	Inflammation Genetics	Niacin Estrogen Genetics	No recommendations at this time

REFERENCES

1. National Cholesterol Education Program. Expert Panel on Detection, Evaluation, and Treatment of High Blood Cholesterol in Adults. Executive Summary of the Third Report of the National Cholesterol Education Program (NCEP) Expert Panel on Detection, Evaluation, and Treatment of High Blood Cholesterol in Adults (Adult Treatment Panel III). JAMA 2001;285:2486-97.
2. Grundy SM, Cleeman JI, Merz CNB, et al. NCEP Report: Implications of Recent Clinical Trials for the National Cholesterol Education Program Adult Treatment Panel III Guidelines. Circulation 2004;110:227-39.
3. Sacco RL, Adams R, Albers G, et al. Guidelines for prevention of stroke in patients with ischemic stroke or transient ischemic attack. Stroke 2006;37:577-617.
4. Kidney Disease Outcomes Quality Initiative (K/DOQI) Group. K/DOQI clinical practice guidelines for management of dyslipidemias in patients with kidney disease. Am J Kidney Dis 2003;41(4 suppl 3):I-IV, S1-S91.
5. Mosca L, Appel LJ, Benjamin EJ, et al. Evidence-based guidelines for cardiovascular disease prevention in women: 2007 update. Circulation 2007;115:1481-501.
6. American Heart Association. AHA/ACC guidelines for secondary prevention for patients with coronary and other atherosclerotic vascular disease: 2006 update. Circulation 2006;113:2363-72.
7. The VAP Cholesterol Test. Available at http://www.atherotech.com/HealthcareProfessionals/default.asp. Accessed July 29, 2008.
8. NMR LipoProfile Test. Available at www.lipoprofile.com/control.cfm?ID=59. Accessed July 28, 2008.
9. Hackman DG, Anand SS. Emerging risk factors for atherosclerotic vascular disease. JAMA 2003;290:932-40.
10. Pearson TA, Mensah GA, Alexander RW, et al. Markers of inflammation and cardiovascular disease, application to clinical and public health practice: a statement for healthcare professionals from the Centers for Disease Control and Prevention and the American Heart Association. Circulation 2003;107:499-511.
11. Wang X, Qin X, Demirtas H, et al. Efficacy of folic acid supplementation in stroke prevention: a meta-analysis. Lancet 2007;369:876-82.
12. Bazzano LA, Reynolds K, Holder KN, He J. Effect of folic acid supplementation on risk of cardiovascular diseases. JAMA 2006;296:2720-6.

REDUCING CORONARY HEART DISEASE RISK THROUGH LIFESTYLE MODIFICATION

Michael B. Bottorff, Pharm.D., FCCP, CLS, and
Carol M. Mason, ARNP, CLS, FAHA

The NCEP ATP III considers primary prevention the arena for affording the greatest opportunity for health care providers to reduce the risk of CHD in the United States today.[1] These recommendations support a clinical approach that targets lifestyle changes that lower cholesterol concentrations and reduce CHD risk, described as therapeutic lifestyle changes (TLCs).

Approaches to TLCs include (1) reduced intake of dietary cholesterol and saturated fats; (2) increased physical activity; and (3) weight control. Complete cessation of smoking (including secondhand smoke) is also part of a comprehensive TLC approach. In addition, lifestyle management is the first step in managing patients with dyslipidemia and is often considered additive to TLC efforts in clinical trials. An approach to TLCs is offered in the NCEP ATP III report as outlined in Figure 1. Epidemiologic studies clearly demonstrate a link between risk factors including tobacco use, hypertension, dyslipidemia and insulin resistance, and CVD.[2]

Consequently, patients must be thoroughly evaluated for lifestyle habits so that the most effective approach to modification can be formulated using the above algorithm. The implementation of an effective patient-specific TLC plan has been shown to involve a team approach involving physicians, nurses, pharmacists, dietitians, and exercise therapists. Various risk reduction therapies are being used in clinics throughout the country and are supported by major medical organizations. The ATP III guidelines support a multidisciplinary approach to help patients and clinicians adhere to national guidelines.

Elevations in LDL-C are a major risk factor for CHD. Studies have shown that the higher the LDL-C, the greater the risk of developing a coronary event or dying from

FIGURE 1. MODEL OF STEPS IN TLCs

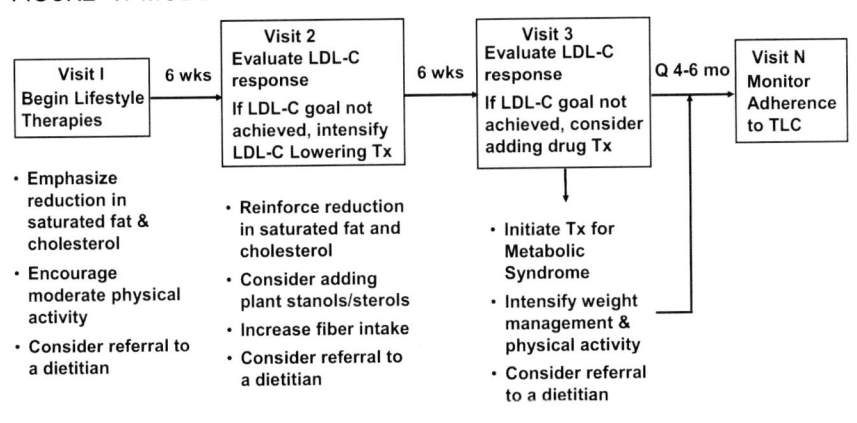

Visit I Begin Lifestyle Therapies	→ 6 wks →	Visit 2 Evaluate LDL-C response If LDL-C goal not achieved, intensify LDL-C Lowering Tx	→ 6 wks →	Visit 3 Evaluate LDL-C response If LDL-C goal not achieved, consider adding drug Tx	→ Q 4-6 mo →	Visit N Monitor Adherence to TLC
• Emphasize reduction in saturated fat & cholesterol • Encourage moderate physical activity • Consider referral to a dietitian		• Reinforce reduction in saturated fat and cholesterol • Consider adding plant stanols/sterols • Increase fiber intake • Consider referral to a dietitian		• Initiate Tx for Metabolic Syndrome • Intensify weight management & physical activity • Consider referral to a dietitian		

LDL-C = low-density lipoprotein cholesterol.
Adapted from Executive Summary of the Third Report of the National Cholesterol Education Program (NCEP) Expert Panel on the Detection, Evaluation, and Treatment of High Blood Cholesterol in Adults (Adult Treatment Panel III). JAMA 2001;285:2486–97.

coronary artery disease. Recent guidelines encourage reductions in LDL-C both with lifestyle and pharmacological therapy, and health care providers today are encouraged to target LDL-C in primary care management.[3]

The importance of TLCs to clinical outcomes was recently highlighted in the INTERHEART study. This study evaluated the attributable risk of modifiable risk factors for myocardial infarction in patients from 52 countries.[4] As seen in Table 1, risk factors for TLC therapy appear to significantly affect clinical outcomes; in many cases, the risk exceeded the targets for drug therapy, such as hypertension and diabetes.

The seven risk factors shown in Table 1 account for 90% of the attributable risk factors for CHD; of interest, these are all modifiable risks that should be included in global risk management in primary care medicine. Of note is that in this European study, the investigators included as a risk factor the "irregular use of fruits and vegetables," a risk not considered in our Western culture. All seven risk factors were consistent in both men and women, across various geographic regions, and among various ethnic groups. The two most important risk factors were smoking and abnormal blood lipids.

REDUCING CHD RISK WITH NUTRITIONAL THERAPY RECOMMENDATIONS

Clinical management of CHD, particularly the risk from abnormal lipids, includes nutritional management that is often best served by a nutritionist trained in managing patients with lipid disorders (see Table 2). Although referral to a trained dietitian is often desirable, a dietician may not always be available, and often, health care provid-

TABLE 1. ASSOCIATION OF RISK FACTORS WITH ACUTE MYOCARDIAL INFARCTION AFTER ADJUSTMENT FOR AGE, SEX, AND GEOGRAPHIC REGION (INTERHEART STUDY)

Risk Factor	Sex	Odds Ratio (99% CI)
Current smoking	W	2.86 (2.36–3.48)
	M	3.05 (2.78–3.33)
Diabetes	W	4.26 (3.51–5.18)
	M	2.67 (2.36–3.02)
Hypertension	W	2.95 (2.57–3.39)
	M	2.32 (2.12–2.53)
Abdominal obesity	W	2.26 (1.90–2.68)
	M	2.24 (2.03–2.47)
Fruits and vegetables	W	0.58 (0.48–0.71)
	M	0.74 (0.66–0.83)
Exercise	W	0.48 (0.39–0.59)
	M	0.77 (0.69–0.85)
ApoB:APOA1 ratio	W	3.43–5.70
	M	3.23–4.38

CI = confidence interval.

Adapted from Yusuf S, Hawken S, Ounpuu S, et al.; INTERHEART study investigators. Effect of potentially modifiable risk factors associated with myocardial infarction in 52 countries (the INTERHEART study): case-control study. Lancet 2004;364:937–52.

ers who work with patients on a daily basis are asked to provide this education. Effective nutritional support can be a valuable tool in risk management. Dietary control of abnormal elevations of LDL-C is a major goal of CVD management. Dietary recommendations that target adults with lipid disorders and weight management issues need to stress the importance of incorporating healthy foods into their diets as opposed to asking patients to go on diets that are not in keeping with their lifestyle or dietary preferences. The American Heart Association (AHA) recommends helping patients avoid inappropriate weight gain during their adult years and helping them better understand the importance of avoiding sedentary activity.

Dietary cholesterol has been shown in clinical studies to raise LDL-C concentrations. A total of 100 mg/day of dietary cholesterol will raise total cholesterol concentrations 2–3 mg/dL. About 70% of total cholesterol is found in the LDL-C subfraction.[3] Diets low in saturated fat have been shown to reduce total cholesterol 8–10%, and when added to a diet that restricts the intake of cholesterol by 200 mg/day, LDL-C may be expected to drop 11–15%.[5]

Asking patients to restrict foods that they have grown up with or to abstain from foods they enjoy may contribute to nonadherence. Instead, we encourage relearning foods that are functionally healthy and helping patients incorporate healthier choices into their diet. The following is an overview of the recommendations for dietary intake by AHA and ATP guidelines. The typical American diet contains excess fat and an increasing number of calories, both of which contribute to dyslipidemia and the obesity epidemic. The NCEP ATP III TLC diet, commonly known as the TLC diet, includes the following recommendations:

TABLE 2. NUTRIENT COMPOSITION OF THE TLC DIET[a]

Nutrient	Recommended Intake
Saturated fat[a]	Less than 7% of total calories
Polyunsaturated fat	Up to 10% of total calories
Monounsaturated fat	Up to 20% of total calories
Total fat	25–35% of total calories
Carbohydrate[b]	50–60% of total calories
Fiber	20–30 g/day
Protein	About 15% of total calories
Cholesterol	Less than 200 mg/day
Total calories (energy)[c]	Balance energy intake and expenditure to maintain desirable body weight/prevent weight gain

[a]Trans fatty acids are another LDL-C–raising fat that should be kept at a low intake.
[b]Carbohydrates should be derived predominantly from foods rich in complex carbohydrates including grains, especially whole grains, fruits, and vegetables.
[c]Daily energy expenditure should include at least moderate physical activity (contributing about 200 kcal/day).

Adapted from Executive Summary of the Third Report of the National Cholesterol Education Program (NCEP) Expert Panel on the Detection, Evaluation, and Treatment of High Blood Cholesterol in Adults (Adult Treatment Panel III). JAMA 2001;285:2486–97.

FAT

Fat can be divided into three categories: saturated, unsaturated, and trans. The number of double bonds in their fatty acid side chains is what differentiates them. Patients need to understand the differences between functionally healthy fats and less healthy fats.

Saturated Fatty Acids

The ATP III guidelines recommend a reduction in saturated fats to less than 7% of an individual's daily calorie consumption. Saturated fats are found in foods containing animal products, such as beef, dairy products, palm and coconut oils, cocoa butter, and hydrogenated fats.[5,6]

All forms of saturated fats increase LDL-C, and patients should be instructed to avoid them or limit their quantity. Table 3 lists examples of common foods and their total and saturated fat amounts.[7]

Unsaturated Fats

Unsaturated fats fall into two categories: polyunsaturated and monounsaturated fats. Polyunsaturated fats lower LDL-C but may also lower the "good" cholesterol, HDL-C. Polyunsaturated fats include omega-3 and omega-6 fatty acids. Dietary sources of omega-6 fatty acids include whole grains, nuts, eggs, and poultry.

Examples of omega-3 fatty acids include flaxseed, fish and fish oils, and eggs, and AHA and ATP III recommendations include two meals of fish per week or 1 g daily of fish oil to reduce the risk of CHD. For patients with elevated triglycerides, increased intake of fish oil may be needed.[8]

TABLE 3. FAT, SATURATED FAT, AND CHOLESTEROL CONTENT OF SELECTED FOODS

Source	Total Fat	Saturated Fat	Cholesterol
Beef rump roast	4.2	1.4	71
Lamb shank	7.8	2.8	78
Pork	11.8	4.1	77
Veal chop	4.9	2.0	93
Liver	4.2	1.6	331
Chicken (light/dark)	3.8/8.3	1.1/2.3	72/79
Salmon	7.0	1.7	54
Lobster	0.5	0.1	61
Shrimp	0.9	0.2	166
Squid	2.4	0.6	400

Adapted from McKenney JM, Hawkins D, eds. Handbook on the Management of Lipid Disorders, 2nd ed. St. Louis, MO: National Pharmacy Cardiovascular Council, 2001:100.

Polyunsaturated Fats

Polyunsaturated fatty acids are found in liquid vegetable oils, such as corn, safflower, and soybean oils, as well as margarine, mayonnaise, and salad dressings. Polyunsaturated fats lower LDL-C but may also lower HDL-C. The ATP III guidelines recommend that adults limit their intake of polyunsaturated fatty acids to no more than 10% of daily calories.[1]

Monounsaturated Fats

Monounsaturated fats include olive oil and canola oil. They tend to lower LDL-C without affecting HDL-C. Recommendations for monounsaturated fats are no more than 20% of total calories, by far the largest recommended portion of daily fat intake. Other good examples of monounsaturated fats are almonds, walnuts, and avocados, all good choices for snacks.

Trans Fats

Trans fats are formed from the partial hydrogenation of vegetable oils, converting them into semisolid fats for use in products such as margarines and oils used in commercial cooking. Trans fats enhance shelf life and stability during deep-frying, making them attractive to the food industry. The U.S. Food and Drug Administration (FDA) now requires nutrition labeling to include trans fat content in recognition of the negative effect of trans fats on total cholesterol, HDL-C, and LDL-C as well as the association of trans fat with high rates of CHD.[9] Because trans fats do not appear to offer any nutritional benefit and are an added risk factor for CHD, many health care sources recommend that they be totally eliminated from the American diet.

PLANT STANOLS/STEROLS

Another recommendation from the NCEP is to increase the intake of plant stanols/sterols (2 g/day) and viscous (soluble) fiber (10–25 g/day) (see Table 4).

TABLE 4. PLANT STANOL/STEROL CONTENT OF FOODS AND SUPPLEMENTS

Food Source	Plant Stanol/Sterol Content	Calories per Serving
Take Control Light Margarine	1700 mg/tbsp (plant sterol esters)	45
Benecol Margarine Light	850 mg/tbsp (plant stanol esters)	50
Minute Maid Heart Wise Orange Juice	1000 mg/8 oz (plant sterols)	110
Health Valley Heart Wise Cereal	400 mg/1 cup (plant sterols)	200
CocoaVia (snack bar)	1500 mg/bar (plant sterol esters)	80
Nature Made CholestOff	450 mg/capsule (plant sterols/stanols)	0.3
BasiChol in flax oil	800 mg/tsp (plant sterol esters)	40

Many vegetables, nuts, seeds, fruits, and grains contain stanols and sterol esters, which decrease cholesterol absorption and thus reduce concentrations of cholesterol in the blood. The average American diet contains few plant stanols/sterols, but when added as a dietary supplement in doses of 2–3 g/day, LDL reductions of up to 15% can be seen. With the addition of these products in recommended doses, LDL-C can be reduced 10–15% with little change in HDL-C or triglycerides.[10,11] Some commercially available plant stanol/sterol food supplements are shown.

DIETARY FIBER

Of the two types of dietary fiber, soluble and insoluble, insoluble fiber has no effect on blood cholesterol concentrations. Soluble (or viscous) fiber may be found in foods such as dried beans, grains, fruits, and vegetables and may lower LDL-C by modest amounts (less than 10%).[12] Despite their minimal effect on lowering LDL-C, high-fiber diets are often rich in vitamins and, together with fewer calories, can still be recommended.

OMEGA-3, OMEGA-6 FATTY ACIDS

Polyunsaturated fatty acids are defined by the first double bond from the methyl end of the molecule. The human body cannot synthesize these fatty acids, and omega-3 fatty acids cannot be transformed from omega-6 fatty acids. The prominent forms of omega-6 fatty acids are arachidonic acid (animal fat) and linoleic acid found in vegetable oils, seeds, and nuts (Table 5). Dietary sources of omega-3 fatty acids (see Table 6) are fish containing eicosapentaenoic acid (EPA) and docosahexaenoic acid (DHA), together with nuts, seeds, and vegetable oils containing α-linoleic acid (ALA), which can be converted to EPA and DHA by a desaturase enzyme in humans. The omega-3 fatty acids have anti-inflammatory and antiplatelet effects and, in high doses, can lower serum triglycerides. Both dietary and supplemental intake of omega-3 fatty acids has

TABLE 5. ALA CONTENT OF SELECTED FOODS

Food	α-Linoleic Acid Content (g/tbsp)
Olive oil	0.1
Walnuts, English	0.7
Soybean oil	0.9
Canola oil	1.3
Walnut oil	1.4
Flaxseeds	2.2
Flaxseed (linseed) oil	8.5

Adapted from Kris-Etherton PM, Harris WS, Appel LJ; for the Nutrition Committee. Fish consumption, fish oil, omega-3 fatty acids and cardiovascular disease. Circulation 2002;106:2747–57.

been shown to reduce cardiovascular morbidity and mortality.[13,14] In 2002, the AHA published a scientific statement on fish consumption, fish oils, and omega-3 fatty acids; key points are summarized in Table 7.[15]

In the past, the safety of omega-3 fatty acid supplementation was questioned, specifically with regard to its effect on bleeding. In a recent report by the National Lipid Association on the safety of nonstatin therapy, a task force concluded that clinical trial evidence does not support an increased risk of bleeding with the supplementation of fish oil therapy.[16]

For ALA ingestion, the FDA has approved a qualified health claim for certain nuts added to a healthy diet. The recommendation supports the consumption of 1.5 oz of nuts/day to help reduce cardiovascular risk. The recommended nuts include tree nuts such as walnuts, almonds, pistachios, and macadamias and may result in LDL-C reductions of up to 15%. Ingestion of larger amounts is to be avoided because of increased caloric value.[17]

EXERCISE

Exercise and increased levels of physical activity can be classified as beneficial for health, fitness, and/or performance. The amount of exercise activity sufficient to attain a health benefit is aerobic exercise of moderate intensity (40–60% of maximal effort) that is undertaken 5–7 days/week for about 30 minutes/day.[18] It is estimated that 61% of American adults do not participate in any form of regular exercise.[19] A sedentary lifestyle has been identified as a leading (and preventable) contributor to all-cause mortality, and there is an inverse and linear relationship between the level of physical activity and death.[20] The benefits of exercise include weight reduction and weight maintenance as well as improvements in parameters defining the metabolic syndrome (such as increasing HDL-C and reducing insulin resistance). However, the exact benefit of overall weight loss in association with an exercise program is often difficult to assess. In many patients, though, who have metabolic syndrome and an enlarged waist circumference (WC), regular aerobic exercise as previously described can improve insulin resistance without altering body mass index (BMI). In several epidemiologic studies, choosing to engage in diverse forms of exercise is associated with lower cardiovascular events.[20]

TABLE 6. SOURCES OF EPA/DHA AND FISH CONSUMPTION

Fish	EPA+DHA Content (per g/3-oz serving of fish or per g/g of oil)	Amount required to provide about 1 g of EPA+DHA per day (per oz of fish or per g of oil)
Sockeye salmon	1.05	2.5 oz
Atlantic salmon (wild)	0.9–1.56	2–3.5
Atlantic salmon (farmed)	1.09–1.83	1.5–2.5
Eastern oysters	095	3 oz
Atlantic cod	0.13	23 oz
Pacific cod	0.24	12.5 oz
Catfish (farmed)	0.15	20

Adapted from Kris-Etherton PM, Harris WS, Appel LJ; for the Nutrition Committee. Fish consumption, fish oil, omega-3 fatty acids and cardiovascular disease. Circulation 2002;106:2747–57.

TABLE 7. AHA RECOMMENDATIONS FOR OMEGA-3 FATTY ACID INTAKE[a]

Patients without documented CHD	Eat a variety of (preferably oily) fish at least twice weekly. Include oils and foods rich in linolenic acid (e.g., flaxseed, canola and soybean oils, walnuts)
Patients with documented CHD	Consume 1 g of EPA+DHA per day, preferably from oily fish. EPA+DHA supplements could be considered in consultation with the physician
Patients needing triglyceride lowering	2–4 g of EPA+DHA per day provided as capsules under a physician's care

[a]*Adapted from Kris-Etherton PM, Harris WS, Appel LJ; for the Nutrition Committee. Fish consumption, fish oil, omega-3 fatty acids and cardiovascular disease. Circulation 2002;106:2747–57.*

Examples of Moderate Physical Activity

- Brisk walking 3–4 mph for 30–60 minutes
- Swimming–laps for 20 minutes
- Running–1.5 miles in 15 minutes
- Bicycling–5 miles in 30 minutes
- Moderate lawn mowing (pushing a power mower) for 30 minutes
- Basketball for 15–20 minutes
- Golf–pulling cart or carrying clubs
- Moderate-intensity social dancing for 30 minutes

SMOKING CESSATION

Thirty percent of the CHD deaths in the United States today are caused by cigarette smoking. Fifty percent of cardiovascular events are attributed to smoking, as is the risk of stroke.[21] The risks from cigarette smoking diminish soon after cessation. The goal of clinical management for patients who smoke should be complete cessation with no exposure to environmental or secondhand smoke. Health care providers should ask patients about tobacco use, assess patient willingness to quit, and assist by developing a plan for quitting, which may include pharmacological intervention (e.g., nicotine

TABLE 8. DEFINITION OF OVERWEIGHT AND OBESITY

	BMI (kg/m^2)	Disease Risk Relative to Normal Weight and WC	
		Men, ≤ 102 cm Women, ≤ 88 cm	Men, > 102 cm Women, > 88 cm
Underweight	< 18.5		
Normal	18.5–24.9		
Overweight	25.0–29.9	Increased	High
Obesity			
Class I	30.0–34.9	High	Very high
Class II	35.0–39.9	Very high	Very high
Class III (extreme obesity)	≥ 40	Extremely high	Extremely high

patches, varenicline).[8] Smoking cessation is often associated with weight gain, and particular attention should be given to preventing this potentially offsetting hazard of smoking cessation.

WEIGHT LOSS

The increase in obesity in the United States and worldwide has often been described as exceeding epidemic proportions. Increased rates of overweight and obesity occur in children, adolescents, and adults; such increases are easily documented in clinical practice according to BMI. Overweight and obesity classifications according to the NCEP ATP III are listed in Table 8.[1]

Overweight and obesity contribute to numerous other medical conditions such as type 2 diabetes mellitus, hypertension, certain cancers, and sleep apnea. Several underlying pathophysiologic conditions drive the risks associated with obesity that contribute to CHD; these include insulin resistance, hypertension, dyslipidemia, increased inflammatory markers, and a prothrombotic state. Obesity, particularly abdominal obesity, is a hallmark feature of metabolic syndrome and is the most common feature found in patients with this cluster of cardiometabolic risk factors.[22] Obesity is also the most common risk factor found in patients with type 2 diabetes mellitus.[23]

The American public is overwhelmed with conflicting and confusing information regarding the best way to lose weight. Fad diets are perceived to be in competition with diets recommended by organizations such as the AHA, which supports a diet low in saturated fat and cholesterol. Several popular diets overly restrict carbohydrates and encourage large intakes of protein and fat (e.g., Atkins, Zone, South Beach). For short periods, patients often do well on these diets but then relapse with even greater increases in overall body weight. Although these so-called fad diets are useful for short-term weight loss in many individuals, they usually result in higher intakes of cholesterol and saturated fat, because the high-protein sources are typically from animal sources. It is difficult to ascertain whether beneficial effects from the fad diets on cardiometabolic risk factors are due to weight loss alone. Recently, the AHA concluded that high-protein diets not reducing total caloric intake do not demonstrate sustained

TABLE 9. HIGH-PROTEIN DIETS AND AHA CRITERIA[24]

AHA Protein Criteria	Sugar Busters	Zone	Atkins
Total diet can be safely implemented during the long term by providing nutrient adequacy and supporting a healthful eating plan to prevent increases in disease risk	No Eliminates many carbohydrates. Discourages eating fruits with meals. Low in calcium, vitamin D, copper, and potassium	No Food must be eaten in required proportions of protein, fat, and carbohydrates. Vegetable portions are very low	No Limited food choices, low in fiber, vitamin D, thiamine, and trace minerals. High in saturated and total fat
Total fat (30%) and saturated fat (10%) are not excessive	Yes 21% total calories, 4% saturated fat per day	Yes 29% total calories, 4% saturated fat per day	No 1st 2 weeks = 53% fat, 26% saturated fat per day
Carbohydrates are not omitted or severely restricted. Minimum of 100 g/day	Yes 114 g/day (52%)	Yes 135 g/day (36%)	No 1st 2 weeks = 28 g/day (5%) Yes Maintenance = 128 g/day
Total protein is not excessive (average 50–100 g/day, proportional 15–20% kcal/day to carbohydrates and fat)	No 71 g/day (27%)	No 127 g/day (34%)	No 1st 2 weeks = 125 g/day (36%); No Maintenance = 110 g/day (24%)

weight loss or improved health.[24] Diets demonstrating sustained weight loss are most often associated with a nutritionally balanced diet plan that considers reductions in daily caloric intake sufficient to produce weight loss of around 1 lb/week (about a 500-kcal/day reduction). Weight management goals seem to be obtained when caloric restriction is coupled with increased physical activity sufficient to produce sustained results. Comparisons of high-protein diets with AHA dietary criteria are shown in Table 9.

SUMMARY

Improving diet and lifestyle is an important component to managing patients at risk of CHD. The ATP and the AHA, together with several other valued American health associations, have published hundreds of guidelines that ask Americans to make major changes in their lifestyle to reduce their risk of death and disability associated with CHD. The primary care arena, whether it is family medicine, internal medicine, or women's health clinics, is often the gatekeeper of CHD prevention, and the health care providers who manage the health of these patients must find creative ways to assist patients in their own disease prevention. It is imperative that, during each visit with their patients, health care providers emphasize the importance of risk reduction, including lowering concentrations of blood cholesterol, reducing risk of hypertension, and quitting smoking. Weight management, preventing type 2 diabetes mellitus, and reducing the risk of CHD are all critically important areas of practice. Elevated concentrations of LDL-C are a major cause of CHD; since its inception in the mid-1980s, the ATP has emphasized the importance of lifestyle management to control this risk factor. All health care providers need to focus on the role of abnormal lipids, and patients should be reminded regularly of the importance of getting to goal.

REFERENCES

1. Executive Summary of the Third Report of the National Cholesterol Education Program (NCEP) Expert Panel on the Detection, Evaluation, and Treatment of High Blood Cholesterol in Adults (Adult Treatment Panel III). JAMA 2001;285:2486-97.
2. Yusuf S, Reddy S, Ounpuu S, Anand S. Global burden of cardiovascular diseases. Part I. General considerations, the epidemiologic transition, risk factors, and impact of urbanization. Circulation 2001;104:2746-53.
3. Fletcher B, Berra K, Ades P, et al. Managing abnormal blood lipids: a collaborative approach. Circulation 2005;112:3184-209.
4. Yusuf S, Hawken S, Ounpuu S, et al.; INTERHEART study investigators. Effect of potentially modifiable risk factors associated with myocardial infarction in 52 countries (the INTERHEART study): case-control study. Lancet 2004;364:937-52.
5. Lichtenstein AH, Appel LJ, Brands M, et al. Diet and lifestyle recommendations revision 2006: a scientific statement from the American Heart Association Nutrition Committee. Circulation 2006;114:82-96.
6. Roussell MA, Kris-Etherton P. Effects of lifestyle interventions on high-density lipoprotein cholesterol levels. J Clin Lipidol 2007;1:65-73.
7. McKenney JM, Hawkins D, eds. Handbook on the Management of Lipid Disorders, 2nd ed. St. Louis, MO: National Pharmacy Cardiovascular Council, 2001:100.
8. Smith SC, Allen J, Blair SN, et al. AHA/ACC guidelines for secondary prevention for patients with coronary and other atherosclerotic vascular disease: 2006 update: endorsed by the National Heart, Lung and Blood Institute. Circulation 2006;113:2363-72.
9. Mozaffarian D, Katan MB, Ascherio A, et al. Trans fatty acids and cardiovascular disease. N Engl J Med 2006;354:1601-13.
10. Amundsen AL, Ose L, Nenseter MS, Ntanios FY. Plant sterol ester-enriched spread lowers plasma total and LDL cholesterol in children with familial hypercholesterolemia. Am J Clin Nutr 2002;76:338-44.
11. Blair SN, Capuzzi DM, Gottlieb SO, et al. Incremental reduction of serum total cholesterol with the addition of plant stanol ester-containing spread to statin therapy. Am J Cardiol 2000;86:46-52.

12. Jenkins DJ, Wolever TM, Rao AV, et al. Effect on blood lipids of very high intakes of fiber in diets low in saturated fat and cholesterol. New Engl J Med 1993;329:21–6.
13. Burr HL, Fehily AM, Gilbert JF, et al. Effects of changes in fat, fish and fibre intakes on death and myocardial reinfarction: Diet and reinfarction trial (DART). Lancet 1989;2:757–61.
14. Dietary supplementation with n-3 polyunsaturated fatty acids and vitamin E after myocardial infarction: Results of the GISSI-Prevenzione trial. Gruppo Italiano per lo Studio della Sopravvivenza nell'Infarto miocardico. Lancet 1999;354:447–55.
15. Kris-Etherton PM, Harris WS, Appel LJ; for the Nutrition Committee. Fish consumption, fish oil, omega-3 fatty acids and cardiovascular disease. Circulation 2002;106:2747–57.
16. Harris WS. Expert opinion: Omega-3 fatty acids and bleeding—cause for concern? Am J Cardiol 2007;99(suppl):44c–46c.
17. Brown D. The FDA considers health claims for nuts. J Am Diet Assoc 2003;103:426.
18. Howly E, Franks D. Health and Fitness Instructors Handbook, 5th ed. Champaign, IL: Huan Kinetics, 2007.
19. American Heart Association. Heart disease and stroke statistics—2007 update. Dallas, TX: AHA, 2007. Available at http://www.americanheart.org/presenter.jhtml?identifier=3055922. Accessed July 23, 2008.
20. Marcus B, Williams DM, Dubbert PM, et al. Physical activity intervention studies: What we know and what we need to know. Circulation 2006;114:2739–52.
21. Ockene IS, Miller NH; for the American Heart Association Task Force on Risk Reduction. Cigarette smoking, cardiovascular disease and stroke. Circulation 1997;96:3243–7.
22. Ford ES, Giles WH, Dietz WH. Prevalence of the metabolic syndrome among US adults. JAMA 2002;287:356–9.
23. Chan JM, Rimm EB, Colditz GA, et al. Obesity, fat distribution and weight gain as risk factors for clinical diabetes in men. Diabetes Care 1994;17:961–9.
24. Jeor ST, Howard BV, Prewitt TE, et al. Dietary protein and weight reduction: A statement for healthcare professionals from the Nutrition Committee of the Council on Nutrition, Physical Activity, and Metabolism of the American Heart Association. Circulation 2001;104:1869–74.

4

METABOLIC SYNDROME

Mark J. Cziraky, Pharm.D., FAHA, CLS,
and Sarah A. Spinler, Pharm.D., FCCP, BCPS (AQ Cardiology)

Cardiometabolic risk has become a commonly used clinical term to describe the cluster of risk factors described as metabolic syndrome that include hypertension, dyslipidemia, diabetes mellitus, and obesity, as well as emerging the risk factors believed to be the result of abdominal obesity such as proinflammatory disease, prothrombotic state, and elevated C-reactive protein (CRP). Using the more "global" description of cardiometabolic risk, all risks related to the metabolic modifications and changes associated with CVD are more accurately described, not just the major traditional parameters of CVD risk. Of importance, the most recent definitions of metabolic syndrome put forth in the AHA and National Heart Lung and Blood Institute (NHLBI) Scientific Statement in 2005 are inclusive of those described in the term cardiometabolic risk; therefore, within this chapter, the term metabolic syndrome will be used in describing cardiometabolic risk.

Cardiovascular risk factors often found concurrently in an individual are also known as risk factor clusters, and the presence of these clusters has been shown to increase the risk of developing CVD and type 2 diabetes mellitus, as well as the incidence of future cardiovascular events and morbidity and mortality related to CVD.[1-10]

Several professional medical organizations have published clinical definitions of metabolic syndrome and include examples of risk factor clustering. Variations in the specific definitions of metabolic syndrome include those put forth by the NCEP in their ATP III report,[11,12] the World Health Organization (WHO) clinical criteria for metabolic syndrome, and the International Diabetes Federation (IDF) (Table 1).[13] These clinical definitions of metabolic syndrome are similar in many ways. The dif-

TABLE 1. PROFESSIONAL ASSOCIATION CRITERIA FOR METABOLIC SYNDROME

	NCEP, AHA-NHLBI	WHO	IDF
Glucose (mg/dL)	Dx: at least 3 FPG > 100 or DM	Dx: ↑ Glucose + 2 IFG/IGT/DM (*must be present*)	Dx: WC + 2 FPG > 100 or DM
BP (mm Hg)	≥ 130/≥ 85 or Tx	≥ 140/≥ 90 or Tx	≥ 130/≥ 85 or Tx
HDL (mg/dL)	M < 40 W < 50	M < 35 W < 40	M < 40 W < 50
Triglycerides (mg/dL)	> 150	> 150	> 150
Central obesity	WC: M > 40 in. F > 35 in.	BMI > 30 kg/m^2, *or* waist-to-hip ratio: M > 0.9, W > 0.85	WC determined based on ethnicity/race (*must be present*)
Other	(none)	Microalbuminuria	(none)
Metabolic syndrome definition	Any three criteria	IFG, IGT, or DM and at least two other criteria	Central obesity plus any two of the remaining four criteria

BP = blood pressure; DM = diabetes mellitus; Dx = diagnosis; FPG = fasting plasma glucose; IDF = International Diabetes Foundation; IFT = immunofluorescence test; IGT = impaired glucose tolerance; Tx = transplantation; WC = waist circumference.

ference between them lies mainly in the identification of primary drivers of the cause of this syndrome and required components to be diagnosed with the metabolic syndrome.[12]

PREVALENCE OF CARDIOMETABOLIC RISK FACTORS IN THE UNITED STATES

As of 2004, 79.4 million adults in the United States had CVD—representing about one in every three adult Americans.[14] In 2004, CVD was identified as an underlying cause of death for 871,517 (36%) of the nearly 2.4 million deaths in the United States and continues to be the leading cause of death in the United States in both men and women.[14,15] The second leading cause of death that same year was cancer, which accounted for 554,643 deaths.[14] Together with its elevated clinical prevalence, CVD is also a serious economic burden in the United States, with health care costs for CHD, stroke, hypertension, and heart failure estimated to be $431.8 billion in 2007.[14]

The global risk of CVD is multifactorial and is affected by the various comorbidities found in the metabolic syndrome, including type 2 diabetes mellitus, hypertension, dyslipidemia, and abdominal obesity. In 2003, an estimated 37% of adults in the United States were identified as having multiple risk factors for CVD (a minimum of two of the following: obesity, diabetes, high blood pressure, high cholesterol, physical inactivity, or current smoking).[16] The age-adjusted prevalence of metabolic syndrome in the United States is 23.7%, and the gender comparison of men (24.0%) and women

FIGURE 1. TRENDS IN PREVALENCE OF OVERWEIGHT AND OBESITY AMONG U.S. ADULTS BY NHANES SURVEY YEAR[19,20]

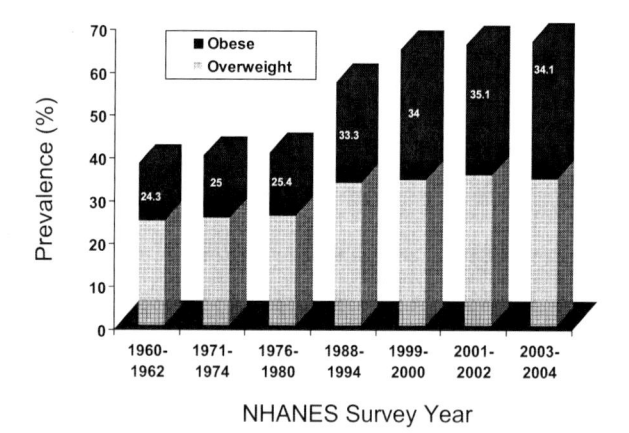

(23.4%) is similar.[17] This breakdown of metabolic syndrome prevalence was lowest (6.7%) among people between 20 and 29 years and was highest (42%) in individuals older than 70 years. With regard to ethnicity, Hispanics had the highest age-adjusted prevalence of metabolic syndrome (31.9%), followed by whites (23.8%) and African Americans (21.6%).[17] Of most concern is the increased prevalence of metabolic syndrome in the U.S. adolescent population. Among a sample of adolescents who were assessed in the National Health and Nutrition Examination Survey (NHANES) III report, the prevalence was 49.7% and 38.7% in the severely obese and moderately obese, respectively.[18]

Currently, the United States is in an "obesity epidemic," which is a significant contributor to the current number of individuals who are classified as having metabolic syndrome. Figure 1 illustrates the obesity rates in the United States and the rate of increase of the past seven reports during 25 years of NHANES follow-up.[19,20] Of importance, obesity and weight gain that occur before the diagnosis of type 2 diabetes mellitus correlate with an increased risk of CVD.[21] The duration of obesity in an individual also has an impact on the development of CVD, with some of the effects becoming evident only after extensive time has elapsed. As an example of this latency and continuous impact of obesity on an individual's CVD risk, patients from the Framingham study who remained in the heaviest weight class during the 26 study-years had the highest risk of CVD consistently during that time frame.[22]

THE PATHOPHYSIOLOGY OF METABOLIC SYNDROME

The development of insulin resistance is a strong predictor of type 2 diabetes mellitus, with abdominal adiposity and inflammation actively involved in several aspects of this process including both the prothrombotic and proinflammatory states found in patients with metabolic syndrome. Several of these cardiovascular risk factors found in patients with metabolic syndrome are caused by the actions of various cytokines (e.g., interleukin-6, tumor necrosis factor α) and adipokines (e.g., adiponectin, leptin) on both muscle tissue and the liver (Figure 2).[23] In addition, the development of cardiovascular risk factors found in metabolic syndrome, including atherogenic dyslipidemia (i.e., low HDL-C and high triglycerides), elevated blood pressure, and increased abdominal obesity, are believed to be related to the elements found in and released by the monocytes and macrophages as well as the adipocytes. Ultimately, this will lead to the development of cardiovascular-related events secondary to type 2 diabetes mellitus and atherosclerosis.

The more recent understanding that adipose tissue is a functional endocrine organ has led to the development of a new proposed mechanism for its impact on CVD development and progression. This reclassification of adipose tissue as an organ has been made mainly because, through the secretion of various peptides such as adipokines, this tissue is able to influence other organ systems, such as the central nervous system, and affect the physiologic function of energy homeostasis.[24] The adipokine leptin acts both locally and systemically, with its actions focused directly on muscle and pancreatic β-cells. Leptin levels are increased through the impact of tumor necrosis factor α, which also affects the uptake and storage of nonesterified fatty acids (NEFAs) and plasma glucose. Adiponectin demonstrates antiatherogenic properties,[24] including

FIGURE 2. PATHOPHYSIOLOGY OF CARDIOMETABOLIC RISK

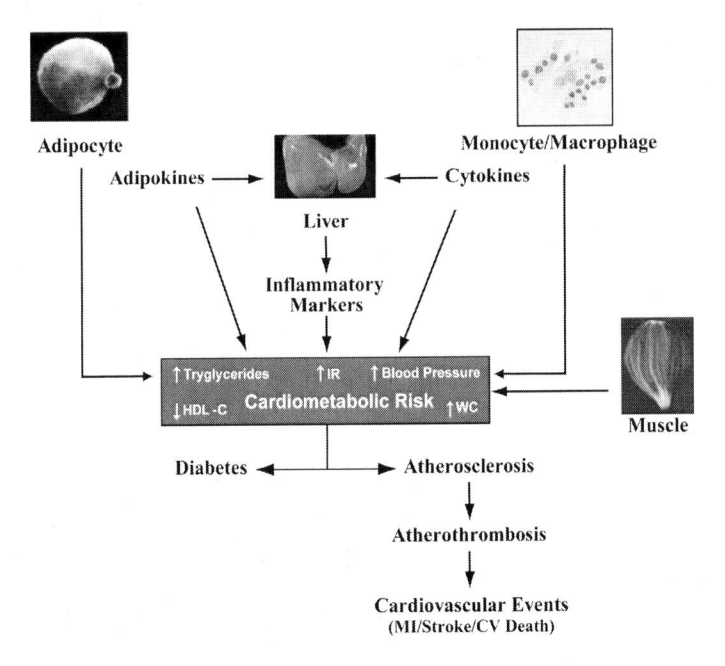

FCV = cardiovascular; IR = insulin resistance; MI = myocardial infarction; WC = waist circumference.

decreased foam cell formation, decreased proliferation, and migration of smooth muscle cells as well as a decreased uptake of oxidized LDL-C by the foam cell. In addition, low levels of adiponectin have been associated with the development of insulin resistance in both animal and human models. An association between adiponectin levels and insulin resistance is present in an inverse fashion whereby increased insulin sensitivity (i.e., decreased insulin resistance) is present with increased adiponectin levels. Plasma insulin and glucose concentrations and percentage of body fat also have an inverse relationship with this peptide.[24]

Elevated free fatty acid concentrations also appear to play a role in the development of insulin resistance. The mechanism by which elevated free fatty acids promote insulin resistance is through an inhibition of the signaling mechanism for insulin, ultimately decreasing the amount of glucose that is cleared from the bloodstream through glucose transport to the muscle.[25] Systemic inflammation, as demonstrated by elevated CRP levels, is positively correlated with the development of insulin resistance and CVD.[1] Cytokines demonstrate inflammatory properties and are elevated in an obese individual—and chronic inflammation is a feature of many patients with metabolic syndrome.

Abdominal obesity is commonly associated with the other major components of the metabolic syndrome including hypertension, diabetes mellitus, and atherogenic dyslipidemia.[24,26-29] As demonstrated in the Nurses Health Study, increased WC is an independent predictor of CHD after adjusting for age.[30] Increased WC is associated with a continuous increase in the risk of CVD in varied race/ethnic populations.[10] During a 13-year follow-up of participants in the Health Professionals Follow-up Study, WC was a strong independent predictor of risk of developing type 2 diabetes mellitus in men.[28] Measuring WC, as the clinically accepted method of assessing abdominal obesity in a patient, is an evaluation that can help predict the presence of multiple cardiometabolic risk factors in an individual (Figure 3).[31] In addition, the excess of adipose tissue present in abdominal obesity releases various amounts of many elements, including plasminogen activator inhibitor type 1 (PAI-1), adiponectin, and NEFAs; each of these can negatively affect the aforementioned associated cardiovascular risk factors.

Insulin resistance is part of the overall cascade of abnormalities associated with metabolic syndrome. Through insulin resistance, glucose concentrations are elevated, which is believed to have a direct impact on the development of various metabolic risk factors.[32] Obesity is commonly associated with atherogenic dyslipidemia, with most obese individuals demonstrating both hyperinsulinemia and decreased insulin sensitivity. This association with and degree of reduced insulin sensitivity, however, is quite variable within the obese population (BMI more than 30 kg/m^2) as well as with overweight individuals (BMI = 25–29.9 kg/m^2). This suggests an inherited component to reduced insulin sensitivity, which can also be supported by the existence of high levels of insulin resistance in world populations where obesity is not prevalent and the population is generally of normal weight (BMI less than 25 kg/m^2).[32]

Insulin resistance may result in increased triglyceride concentrations and reduced HDL concentrations through several mechanisms. Insulin inhibits hepatic VLDL secretion, and insulin resistance results in a loss of this inhibition, leading to an

FIGURE 3. MEASURING WAIST CIRCUMFERENCE[31]

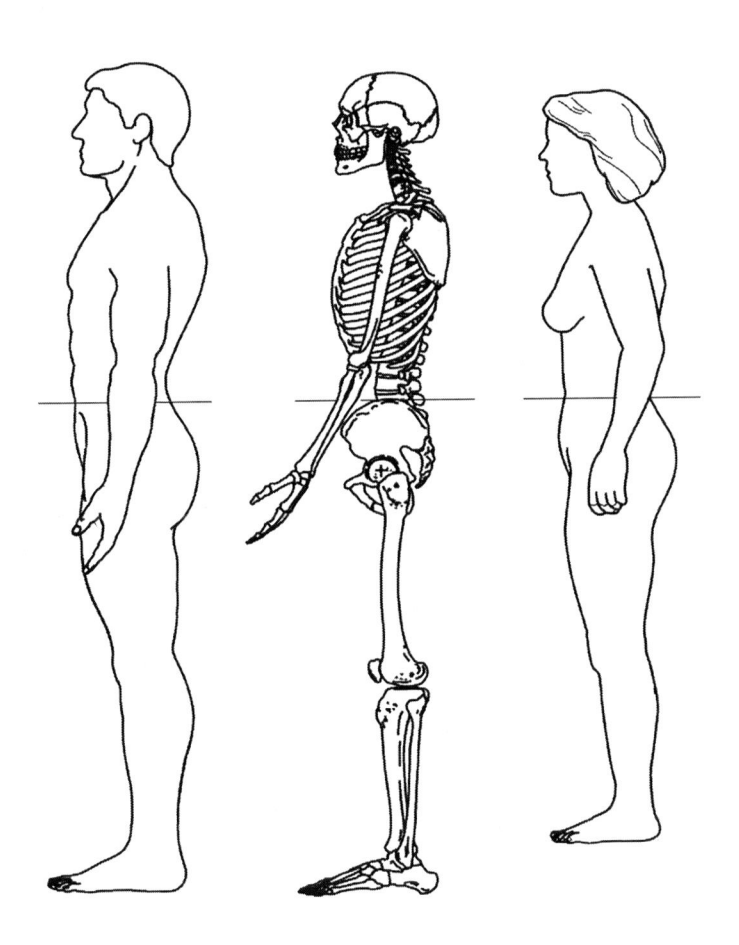

Instructions for Measuring Waist Circumference, according to NHANES III Protocol. To define the level at which waist circumference is measured, a bony landmark is first located and marked. The subject stands, and the examiner, positioned at the right of the subject, palpates the upper hip bone to locate the right iliac crest. Just above the uppermost lateral border of the right iliac crest, a horizontal mark is drawn and then crossed with a vertical mark on the midaxillary line. The measuring tape is placed in a horizontal plane around the abdomen at the level of this marked point on the right side of the trunk. The plane of the tape is parallel to the floor, and the tape is snug but does not compress the skin. The measurement is made at a normal minimal respiration (see Figure 5). REF: U.S. Department of Health and Human Services, PHS. HANES III anthropometric procedures video. U.S. Government Printing Office Stock 017-022-01335-5. Washington, DC: U.S. GPO, Public Health Service, 1996:538.

Reproduced with permission. Available at www.ncbi.nlm.nih.gov/books/ bv.fcgi?rid=obesityfiggrp.237.

FIGURE 4. EFFECTS OF INSULIN RESISTANCE ON VLDL PRODUCTION AND FUNCTION IN OBESITY[33]

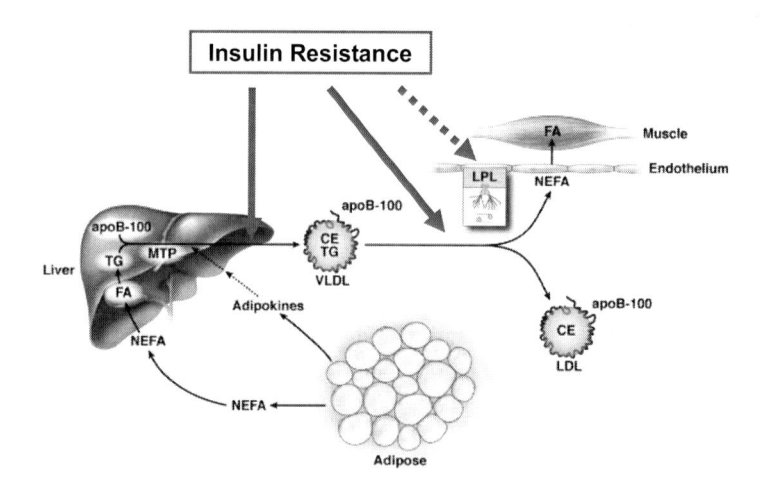

apoB = apolipoprotein B; FA = fatty acids; LDL = low-density lipoprotein cholesterol; LPL = lipoprotein lipase; MTP = microsomal triglyceride transfer protein; NEFA = nonesterified fatty acids; TG = triglycerides; VLDL = very low-density lipoprotein cholesterol.

Adapted with permission from reference 33: Bamba V, Rader D. Obesity and atherogenic dyslipidemia. Gastroenterology 2007;132:2181–91.

increased production of triglycerides (Figure 4).[33] One proposed mechanism for insulin resistance is that microsomal triglyceride transfer protein is up-regulated. This results in an increased formation of VLDLs by the endoplasmic reticulum as well as an increased transfer of mature lipid particles to apolipoprotein B. Insulin resistance also impairs lipoprotein lipase activity, thus decreasing HDL CTEP transfer of cholesterol esters from HDL to triglyceride-rich lipoproteins in exchange for triglycerides (Figure 5). In addition, increased triglyceride-rich lipoprotein activity results in the production of triglyceride-rich HDL particles that are then hydrolyzed by HL to smaller HDL particles. Thus, HDL concentrations are reduced, and those that are produced are dense and rich in triglycerides. Finally, insulin resistance is thought to directly increase the activity of CTEP and HL.[33]

Hyperinsulinemia is also believed to lead to an atherogenic dyslipidemia profile in individuals with insulin resistance. There is an increase in the release of VLDLs, causing an elevation in triglycerides as well as an increase in blood pressure through an unidentified mechanism. Insulin-resistant muscle tissue existing in an environment with hyperinsulinemia is overloaded with lipids secondary to increased plasma NEFA concentrations, leading to fatty liver development through the diversion of this excess lipid from the muscle tissue to the liver.[33]

FIGURE 5. EFFECTS OF INSULIN RESISTANCE ON PRODUCTION AND FUNCTION OF HDL[33]

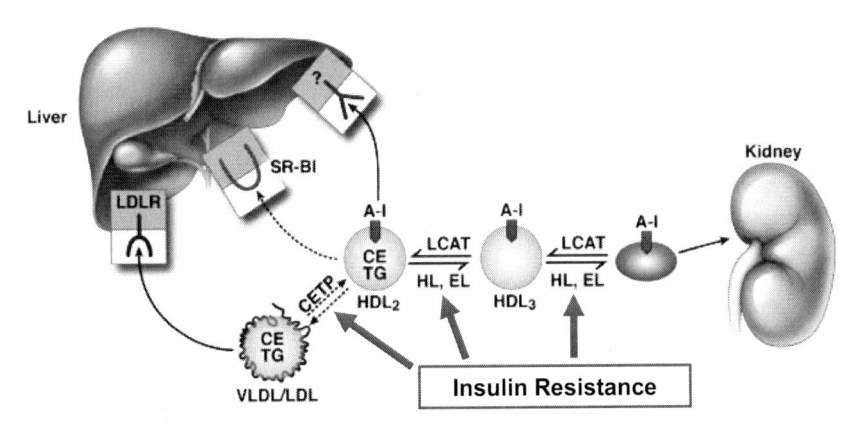

CE = cholesterol esters; CETP = cholesterol ester transfer protein; EL = endothelial lipase; HDL = high-density lipoprotein cholesterol; HL = hepatic lipase; LCAT = lecithin cholesterol acyltransferase; LDLR = low-density lipoprotein cholesterol receptor; SR-BI = hepatic scavenger receptor class B type I; VLDL = very low-density lipoprotein cholesterol.

Adapted with permission from reference 33: Bamba V, Rader D. Obesity and atherogenic dyslipidemia. Gastroenterology 2007;132:2181–91.

IDENTIFICATION AND MANAGEMENT OF METABOLIC SYNDROME IN PATIENTS

In 2005, the AHA and NHLBI convened a special collaborative conference to focus on the issues surrounding the definition of metabolic syndrome, specifically:

1. Major clinical outcomes
2. Metabolic components
3. Pathogenesis
4. Clinical criteria for diagnosis
5. Risk of clinical outcomes
6. Therapeutic interventions

The primary goal for managing metabolic syndrome described in the Scientific Statement developed from this conference was to reduce the risk of developing atherosclerotic CVD.[34] The NCEP ATP III criteria remain the most commonly used in the United States for diagnosing metabolic syndrome and will therefore be the criteria used in this chapter for this definition. Six main components of the metabolic syndrome were identified in the NCEP ATP III report:

1. Abdominal obesity
2. Atherogenic dyslipidemia
3. Elevated blood pressure
4. Insulin resistance with or without glucose intolerance
5. Proinflammatory state
6. Prothrombotic state

Within these components are several specific risk factors that the NCEP ATP III described as underlying, major, or emerging risk factors. Table 2 categorizes the specific risk factors into these three subgroups, and these risk factors are used during the evaluation of patients and their history. They assist the clinician in the diagnosis of the metabolic syndrome. Through individual patient evaluation, the metabolic syndrome exists in an individual when a minimum of three of these five risk factors are present.[12]

A stepwise approach to the identification and management of patients with metabolic syndrome has been described (Table 3).[35] The initial steps of assessing the risk of an individual with initial therapies targeted the major individual components of metabolic syndrome including hypertension, dyslipidemia, and type 2 diabetes mellitus. Non-pharmacological therapy such as lifestyle modifications, including weight reduction, smoking cessation, and increased physical activity, together with appropriate dietary modifications, should be implemented. Specifically, diets that institute low amounts of saturated fats, trans fats, cholesterol, and both simple sugars and sodium should be implemented. Increased consumption of fruits and vegetables, whole-grain bread products, and fish should be recommended.[11,34] If non-pharmacological methods for managing individual cardiovascular risk factors in these individuals are not sufficient, medication therapy is warranted and should be initiated in a consistent fashion with evidence-based guidelines for hypertension, dyslipidemia, and diabetes mellitus (Table 3).[35] Currently, no specific antihypertensive class or dyslipidemic drug has been identified through research as being more efficacious when used in patients with metabolic syndrome. Specifically, these therapies should be targeted at achieving appropriate goal levels for blood pressure, LDL-C, blood pressure, glucose, and hemoglobin A1c. Key points in applying pharmacological therapy are to target the complete metabolic profile rather than one specific risk factor at a time and not to overlook the impact of these collective therapies on reducing abdominal obesity. Currently, there is no specific drug approved in the United States recommended for people with metabolic syndrome independent of those that are most appropriate for specific abnormal cardiovascular risk factors found in these individuals.

Prevention of the progression of specific risk factors such as type 2 diabetes mellitus or dyslipidemia is an important approach to the pharmacological management of these individuals. The efficacy of treating patients with type 2 diabetes mellitus but without overt CHD was recently investigated in the Anglo-Scandinavian Cardiac Outcomes Trial—Lipid-Lowering Arm.[36] In this study, 2532 patients received a diagnosis of type 2 diabetes mellitus and had three or more CVD risk factors (hypertension, age older than 55, smoking, dyslipidemia, etc.) but no history of CVD at baseline; they were randomized to receive 10 mg/day of placebo or atorvastatin. Patients in the atorvastatin group experienced a 23% reduction (hazard ratio 0.77, 95% CI, 0.61–0.98, p=0.036) in cardiovascular events during the 3.3-year study.[36] Similarly, in the recent Collaborative Atorvastatin Diabetes Study, a 36% reduction in cardiovascular events was observed in patients with type 2 diabetes mellitus but no history of CVD who were treated with 10 mg/day of atorvastatin versus placebo.[37] These studies underscore the need to institute a more aggressive management of the CVD risk in patients with type 2 diabetes mellitus.

TABLE 2. AHA-NHLBI IDENTIFICATION OF PATIENTS WITH METABOLIC SYNDROME

Measure (3 of 5 constitutes diagnosis)	Categoric Cut Points
Elevated WC (inches)	≥ 40 (men), ≥ 35 (women)
Elevated triglycerides (mg/dL)	≥ 150 or drug therapy
Reduced HDL-C (mg/dL)	< 40 (men), < 50 (women) or drug therapy
Elevated BP (mm Hg)	≥ 130/85 or drug therapy
Elevated fasting glucose (mg/dL)	≥ 100 or drug therapy

BP = blood pressure; WC = waist circumference.

TABLE 3. AHA-NHLBI STEPWISE APPROACH TO MANAGING METABOLIC SYNDROME

	Description
Step 1:	
Assess risk	For patients without atherosclerotic CVD, estimate 10-year risk of CHD using the Framingham Risk Score
	High-risk patients have atherosclerotic CVD or diabetes with other CHD risk factors: • High risk is 20% • Moderately high risk is 10–20% • Low-moderate risk is 10%
Step 2:	
Manage underlying risk factors	Abdominal obesity: • Calculate BMI • For 7–10% weight loss in 6–12 months, decrease calories by 500–1000 kcal/day • Increase physical activity to 30 minutes 5 days/week • Consider use of weight loss drugs in certain patients • Consider bariatric surgery in certain patients
	Physical inactivity: • Advise 60 minutes/day of continuous or intermittent aerobic activity • Medically supervise patients with recent acute coronary syndrome or percutaneous coronary intervention
	Atherogenic and diabetogenic diet: • Consider exercise stress testing in high-risk individuals • For weight control, decrease total calories • Advise a diet low in saturated fats, trans fats, cholesterol, sodium, and simple sugars • Recommend consumption of fruits, vegetables, whole grains, and fish • According to the ATP III, 25–35% of calories should be from total fat
Step 3:	
Measure fasting plasma glucose concentrations to identify patients with diabetes and prediabetes[a]	Impaired fasting glucose = 100–125 mg/dL
	Consider an oral glucose tolerance test: • Impaired glucose tolerance = 2-hour glucose concentration[3] 140 and < 200 mg/dL
Step 3a:	
Treat patients with diabetes and prediabetes[a]	For diabetes, hemoglobin A1c should be < 6%
	For prediabetes, delay or prevent onset of diabetes with lifestyle modifications

(Continued)

TABLE 3. AHA-NHLBI STEPWISE APPROACH TO MANAGING METABOLIC SYNDROME (Continued)

Step 4:	
Measure fasting lipid panel; treat atherogenic dyslipidemia	LDL concentration primary target: • For CHD and CHD risk equivalents, goal is < 100 mg/dL • For two or more risk factors, goal is < 130 mg/dL
	For non-HDL concentration, < 130 mg/dL is secondary target Triglyceride concentration: • For no or one risk factor, < 150 mg/dL is the goal • Treat regardless of LDL concentration if > 500 mg/dL
Step 5:	
Measure and treat elevated blood pressure	JNC 7 targets: • < 140/90 mm Hg • < 130/80 mm Hg for patients with diabetes, chronic renal disease (stage 5)
Step 6:	
Consider aspirin therapy	To reduce the prothrombotic state: • 81–325 mg/day for established atherosclerotic CVD • 81 mg/day for a Framingham Risk Score of 10% • 75–162 mg/day for patients with diabetes type 1 or 2 who are > 40 or those > 20 with other CVD risk factors
Optional: Measure CRP level	Emerging target for assessment and treatment: • If concentration is > 3 mg/L, consider lifestyle modification

[a]Diabetes mellitus is defined as a group of metabolic diseases characterized by hyperglycemia resulting from insufficient insulin secretion, insulin action, or both. Prediabetes is the American Diabetes Association term for impaired fasting glucose or impaired glucose tolerance.

ATP III = Third Report of the National Cholesterol Education Program Expert Panel on Detection, Evaluation, and Treatment of High Blood Cholesterol in Adults; BMI = body mass index; CHD = coronary heart disease; CVD = cardiovascular disease; HDL = high-density lipoprotein cholesterol; JNC 7 = Seventh Report of the Joint National Committee on Prevention, Detection, Evaluation, and Treatment of High Blood Pressure; LDL = low-density lipoprotein cholesterol.

It is estimated that as many as 12 million people in the United States are candidates for interventions to prevent the onset of type 2 diabetes mellitus.[38] In addition to the Finnish Diabetes Prevention Study of lifestyle modification,[39,40] two clinical trials have shown that aggressive intervention with drugs can prevent or delay the development of type 2 diabetes mellitus in patients with prediabetes. The STOP-NIDDM trial demonstrated that treatment with acarbose resulted in a 25% reduction in the incidence of type 2 diabetes mellitus in patients with elevated fasting plasma glucose or impaired glucose tolerance.[41] Similarly, intervention with metformin or lifestyle therapy significantly reduced the progression to type 2 diabetes mellitus by 31% and 58%, respectively.[42]

Atherogenic dyslipidemia (e.g., low HDL and elevated triglycerides) is the common lipid abnormality found in individuals with metabolic syndrome. Low-density lipoprotein cholesterol continues to be the primary target for therapy; even in the patient with metabolic syndrome, statins are considered first-line therapy, and a recent subanalysis of major statin trials demonstrated that these agents reduce CVD events in patients who were defined as having metabolic syndrome (Treat to New Targets Trial reanalysis). Not until the LDL-C concentration is controlled should the other abnormalities in the atherogenic dyslipidemic profile be targeted with pharmacological therapy. In patients with triglyceride concentrations greater than 200 mg/dL, non–HDL-C becomes the next target of treatment after the LDL-C target is reached (Table 3).[35] In patients with triglycerides greater than 500 mg/dL, triglyceride-lowering therapy should be implemented immediately to prevent the potential pancreatitis that can develop subsequent to the significantly elevated triglyceride concentrations. A third aim for cholesterol management in the patient with metabolic syndrome is the elevation of low HDL-C concentrations with either fibrate or niacin therapy.

Fibrates have been shown to lower triglyceride concentrations and raise HDL-C concentrations. The combination of a fibrate with a statin demonstrates a positive combined impact on the entire lipid panel.[11,43] When statins and fibrates are used in combination, clinical trials demonstrate a greater improvement in the lipid panel abnormalities, but the effect of these improved lipid values on subsequent cardiovascular events has not yet been researched. Although the impact is favorable on the specific lipid fractions that are often abnormal in patients with metabolic syndrome, insufficient clinical trials have been conducted to evaluate the efficacy of the combination of these agents compared with either agent alone in the patient with metabolic syndrome. If non–HDL-C concentrations remain elevated in individuals after LDL-C targets are attained, the clinician can either intensify the LDL-C–lowering regimen, which will also positively affect the individual's non–HDL-C, or add a triglyceride-lowering therapy in combination with the statin therapy. Fenofibrate causes less severe myopathy when used in combination with a statin than does gemfibrozil, and because of this, it is the preferred fibrate to use in combination with a statin. Niacin therapy also decreases non–HDL-C concentrations and the risk of subsequent events in patients with metabolic syndrome/type 2 diabetes mellitus. The Action to Control Cardiovascular Risk in Diabetes trial is designed to test intensive glucose control (less than 6.0 mg/dL), control of atherogenic dyslipidemia through fibrate therapy, and intensive blood pressure control (systolic blood pressure less than 120 mm Hg) on the impact of major cardiovascular events in patients with type 2 diabetes mellitus.[44] The arm of intensive glucose control in this randomized trial of 10,251 patients was stopped early at a median of 3.5 years of follow-up secondary to increased mortality and higher frequency of hypoglycemia.[45] The arms of intensive blood pressure control and the addition of fenofibrate versus placebo to simvastatin therapy are ongoing with a plan to report results in 2010.[44,45]

Current hypertension guidelines identify a blood pressure target of less than 140/90 mm Hg for individuals without concomitant type 2 diabetes mellitus or CKD. With either of these two comorbidities present, the blood pressure target is reduced to less than 130/80 mm Hg. Lifestyle modifications, including increased exercise and dietary therapy, can have a significant impact on an individual's blood pressure. Moreover, diets such as those researched in the Dietary Approaches to Stop Hypertension research initiative have demonstrated a positive impact on blood pressure through the increased consumption of fruits and vegetables and low-fat dairy products together with sodium and alcohol restriction. Because type 2 diabetes mellitus is so often found in patients with metabolic syndrome, many clinicians use angiotensin-converting enzyme inhibitors or angiotensin receptor blockers as first-line therapy—especially if type 2 diabetes mellitus or CKD is already present.[46]

Treatment of obesity includes both pharmacological and non-pharmacological interventions, depending on the individual patient and his or her response to a stepwise approach to weight reduction. Initial therapy should consist of a weight reduction program that includes both a caloric decrease and an increase in physical activity. Many times, the latter depends on the ability of the individual; however, caloric-restricted nutritional therapy should be implemented in all obese individuals. Weight reduction decreases PAI-1 levels in individuals, and the pro-inflammatory state appears to be positively affected, as shown by a reduction in CRP levels. In addition, weight reduction has been associated with reductions in total cholesterol and LDL-C, increased HDL-C concentrations, and decreased blood pressure. In one study, a weight loss of 5% or more during a 3-year period resulted in a significant improvement in concentrations of total cholesterol and LDL-C.[47] The Diabetes Prevention Program Research Group study of individuals with impaired glucose tolerance found that aggressive lifestyle intervention with a goal of 7% weight reduction resulted in significant improvements in triglyceride and HDL-C concentrations compared with placebo or metformin.[48] Insulin resistance is reduced in individuals who have lost weight, which, in turn, leads to a reduction in glucose concentrations in these patients. Weight reductions of 10% or more during a 1-year period in obese women age 25–44 produced a significant improvement in their fasting insulin and glucose concentrations and resulted in reductions in the levels of cytokines associated with a proinflammatory state.[49] In the Finnish Diabetes Prevention Study, weight loss was the best predictor of protection from the development of type 2 diabetes mellitus.[39,40]

Weight loss drugs can be considered when other strategies targeted at abdominal obesity have failed to produce the desired results. These agents are indicated for obese patients with an initial BMI of 30 kg/m^2 or greater or 27 kg/m^2 or greater with the presence of other cardiovascular risk factors. A variety of drugs are used for weight loss, including the lipase inhibitor orlistat (Xenical), and various appetite suppressants, including sibutramine (Meridia and Reductil), bupropion, diethylpropion, fluoxetine, and phentermine.[50] A meta-analysis performed with research studies including these agents found that weight loss at 12 months of therapy was greatest for sibutramine (4.45 kg), followed by fluoxetine (3.15 kg), bupropion (2.8 kg), and orlistat (2.89 kg).[51] Six-month data are available for phentermine (3.6 kg) and diethylpropion (3.0 kg), but importantly, there are no data to determine if one drug is better because of the lack of comparative head-to-head trials.[51]

Sibutramine was demonstrated in the Sibutramine Trial in Obesity Reduction and Maintenance (STORM) to increase weight loss in conjunction with a low-calorie diet greater than placebo throughout the 24-month trial period. Of importance in this trial of 605 obese patents, the group receiving sibutramine lost a mean of about 3.5 inches of WC during the trial period. There was a reduction in triglyceride concentrations of 25% and an increase in HDL-C concentrations of 20% in the group receiving sibutramine. The more common adverse effects associated with sibutramine include headache, constipation, insomnia, tachycardia, and elevated blood pressure. In the STORM study, the latter two adverse events were seen in 2.6% and 2.1% of patients, respectively. It is recommended that blood pressure and heart rate be followed closely during the initiation of treatment because these adverse events seem to occur during the early part of therapy.[52]

An investigative agent, rimonabant, which targets the endocannabinoid system

(ECS), is currently under review at the FDA and under review for the treatment and management of obesity. The ECS is composed of the cannabinoid (CB) receptors (CB1 and CB2), the endogenous ligands that bind to these receptors (anandamide and 2-arachidonyl glycerol), and the enzymes that control the synthesis and degradation of the ligands.[53] Studies indicate that the ECS plays an important role in regulating energy balance, food intake, lipogenesis, and appetite. Cannabinoid receptors are located in the areas of the central nervous system that control appetite and energy balance in peripheral nerves and in peripheral organs that are important for metabolic control. Normally, the ECS acts in an anabolic fashion to promote food intake and energy storage. However, there is evidence to suggest that in obese individuals, the ECS is overactive, which may play a role in their susceptibility to weight gain. In clinical trials of overweight or obese individuals, rimonabant has demonstrated weight reduction, increased HDL-C, decreased triglycerides, and improved measures of insulin resistance as well as a reduction in the percentage of the study population meeting the AHA-NHLBI definition of metabolic syndrome.[54-56]

Another therapeutic target for drug development is the adipose-derived hormone adiponectin. Low levels of adiponectin are found in obese individuals and in individuals with obesity-related diseases such as type 2 diabetes mellitus, CHD, and hypertension. Low levels of adiponectin are associated with inflammatory states and insulin resistance. Blockade of the ECS results in increased levels of adiponectin and a reduction in CRP in overweight individuals and has the potential for a positive therapeutic impact on cardiometabolic risk factors.[54-56]

Low-dose aspirin therapy is used in patients with metabolic syndrome to combat the prothrombotic state that exists in these individuals. Because no drugs are currently available that target fibrinogen and PAI-1, antiplatelet therapy is used to help combat this prothrombotic state. Aspirin therapy, together with statin drugs, has also been demonstrated to reduce CRP levels in these individuals.

CONCLUSION

Although the exact etiology of metabolic syndrome remains unknown, the root cause is most likely a multitude of many factors including a complex set of interactions that occur between both environmental and genetic factors.[57] Currently available management strategies target individual risk factors, with goals of therapy similar to those for patients without metabolic syndrome.

REFERENCES

1. Grundy SM. Metabolic syndrome: connecting and reconciling cardiovascular and diabetes worlds. J Am Coll Cardiol 2006;47:1093–100.
2. Girman CJ, Dekker JM, Rhodes T, et al. An exploratory analysis of criteria for the metabolic syndrome and its prediction of long-term cardiovascular outcomes: the Hoorn study. Am J Epidemiol 2005;162:438–47.
3. Isomaa B, Almgren P, Tuomi T, et al. Cardiovascular morbidity and mortality associated with the metabolic syndrome. Diabetes Care 2001;24:683–9.
4. Laaksonen DE, Lakka HM, Niskanen LK, et al. Metabolic syndrome and development of diabetes mellitus: application and validation of recently suggested definitions of the metabolic syndrome in a prospective cohort study. Am J Epidemiol 2002;156:1070–7.
5. Lakka HM, Laaksonen DE, Lakka TA, et al. The metabolic syndrome and total and cardiovascular disease mortality in middle-aged men. JAMA 2002;288:2709–16.
6. McNeill AM, Rosamond WD, Girman CJ, et al. The metabolic syndrome and 11-year risk of incident cardiovascular disease in the atherosclerosis risk in communities study. Diabetes Care 2005;28:385–90.
7. Port SC, Goodarzi MO, Boyle NG, et al. Blood glucose: a strong risk factor for mortality in nondiabetic patients with cardiovascular disease. Am Heart J 2005;150:209–14.
8. Stamler J, Vaccaro O, Neaton JD, et al. Diabetes, other risk factors, and 12-yr cardiovascular mortality for men screened in the Multiple Risk Factor Intervention Trial. Diabetes Care 1993;16:434–44.
9. Turner RC, Millns H, Neil HA, et al. Risk factors for coronary artery disease in non-insulin dependent diabetes mellitus: United Kingdom Prospective Diabetes Study (UKPDS: 23). BMJ 1998;316:823–8.
10. Zhu S, Heymsfield SB, Toyoshima H, et al. Race-ethnicity–specific waist circumference cutoffs for identifying cardiovascular disease risk factors. Am J Clin Nutr 2005;81:409–15.
11. Third Report of the National Cholesterol Education Program (NCEP) Expert Panel on Detection, Evaluation, and Treatment of High Blood Cholesterol in Adults (Adult Treatment Panel III) final report. Circulation 2002;106:3143–421.
12. Grundy SM, Brewer HB Jr, Cleeman JI, et al. Definition of metabolic syndrome: Report of the National Heart, Lung, and Blood Institute/American Heart Association conference on scientific issues related to definition. Circulation 2004;109:433–8.
13. Alberti KG, Zimmet PZ. Definition, diagnosis and classification of diabetes mellitus and its complications. Part 1. Diagnosis and classification of diabetes mellitus provisional report of a WHO consultation. Diabet Med 1998;15:539–53.
14. Thom T, Haase N, Rosamond W, et al. Heart disease and stroke statistics—2006 update: a report from the American Heart Association Statistics Committee and Stroke Statistics Subcommittee. Circulation 2006;113:e85–151.
15. Minino AM, Heron MP, Smith BL. Deaths: preliminary data for 2004. Natl Vital Stat Rep 2006;54:1.

16. Centers for Disease Control and Prevention. Racial/ethnic and socioeconomic disparities in multiple risk factors for heart disease and stroke—United States, 2003. Morb Mortal Wkly Rep 2005;54:113–7.
17. Ford ES, Giles WH, Dietz WH. Prevalence of the metabolic syndrome among US adults. JAMA 2002;287:356–9.
18. Weiss R, Dziura J, Burgert TS, et al. Obesity and the metabolic syndrome in children and adolescents. N Engl J Med 2003;350:2362–74.
19. Ogden CL, Carroll MD, Curtin LR, et al. Prevalence of overweight and obesity in the United States, 1999–2004. JAMA 2006;295:1549–55.
20. Flegal KM, Carroll MD, Ogden CL, et al. Prevalence of overweight and obesity in the United States, 1999–2004. JAMA 2002;288:1723–7.
21. Cho E, Manson JE, Stampfer MJ, et al. A prospective study of obesity and risk of coronary heart disease among diabetic women. Diabetes Care 2002;25:1142–8.
22. Hubert HB, Feinleib M, McNamara PM, et al. Obesity as an independent risk factor for cardiovascular disease: a 26-year follow-up of participants in the Framingham Heart Study. Circulation 1983;67:968–77.
23. Reilly MP, Rader DJ. The metabolic syndrome: more than the sum of its parts? Circulation 2003;108:1546–51.
24. National Institutes of Health. Clinical guidelines on the identification, evaluation, and treatment of overweight and obesity in adults—the evidence report. Obes Res 1998;6:51S–209.
25. Boden G, Shulman GI. Free fatty acids in obesity and type 2 diabetes: defining their role in the development of insulin resistance and ß-cell dysfunction. Eur J Clin Invest 2007;32:14.
26. Carr DB, Utzschneider KM, Hull RL, et al. Intra-abdominal fat is a major determinant of the National Cholesterol Education Program Adult Treatment Panel III criteria for the metabolic syndrome. Diabetes 2004;53:2087–94.
27. Lakka HM, Lakka TA, Tuomilehto J, et al. Abdominal obesity is associated with increased risk of acute coronary events in men. Eur Heart J 2002;23:706–13.
28. Wang Y, Rimm EB, Stampfer MJ, et al. Comparison of abdominal adiposity and overall obesity in predicting risk of type 2 diabetes among men. Am J Clin Nutr 2005;81:555–63.
29. Zhu S, Wang Z, Heshka S, et al. Waist circumference and obesity-associated risk factors among whites in the third National Health and Nutrition Examination Survey: clinical action thresholds. Am J Clin Nutr 2002;76:743–9.
30. Rexrode KM, Carey VJ, Hennekens CH, et al. Abdominal adiposity and coronary heart disease in women. JAMA 1998;280:1843–8.
31. U.S. Department of Health and Human Services PHI. Anthropometric procedures video. 017-022-01335-5, Washington, DC: U.S. GPO, Public Health Service, 1996:538.
32. Ferrannini E, Haffer SM, Mitchell BD, Stern MP. Hyperinsulinaemia: the key feature of a cardiovascular and metabolic syndrome. Diabetologia 1991;34:416–22.
33. Bamba V, Rader D. Obesity and atherogenic dyslipidemia. Gastroenterology 2007;132:2181–91.
34. Grundy SM, Cleeman JI, Daniels SR, et al. Diagnosis and management of the

metabolic syndrome. An American Heart Association/National Heart, Lung, and Blood Institute Scientific Statement. Executive summary. Cardiol Rev 2005;13:322–7.
35. Spinler SA. Challenges associated with metabolic syndrome. Pharmacotherapy 2006;26:209S–17S.
36. Sever PS, Dahlof B, Poulter NR, et al. Prevention of coronary and stroke events with atorvastatin in hypertensive patients who have average or lower-than-average cholesterol concentrations, in the Anglo-Scandinavian Cardiac Outcomes Trial–Lipid Lowering Arm (ASCOT-LLA): a multicentre randomised controlled trial. Lancet 2003;361:1149–58.
37. Colhoun HM, Betteridge DJ, Durrington PN, et al. Primary prevention of cardiovascular disease with atorvastatin in type 2 diabetes in the Collaborative Atorvastatin Diabetes Study (CARDS): multicentre randomised placebo-controlled trial. Lancet 2004;364:685–96.
38. Benjamin SM, Valdez R, Geiss LS, et al. Estimated number of adults with prediabetes in the US in 2000: opportunities for prevention. Diabetes Care 2003;26:645–9.
39. Lindstrom J, Ilanne-Parikka P, Peltonen M, et al. Sustained reduction in the incidence of type 2 diabetes by lifestyle intervention: follow-up of the Finnish Diabetes Prevention Study. Lancet 2006;368:1673–9.
40. Tuomilehto J, Lindstrom J, Eriksson JG, et al. Prevention of type 2 diabetes mellitus by changes in lifestyle among subjects with impaired glucose tolerance. N Engl J Med 2001;344:1343–50.
41. Chiasson JL, Josse RG, Gomis R, et al. Acarbose for prevention of type 2 diabetes mellitus: the STOP-NIDDM randomised trial. Lancet 2002;359:2072–7.
42. Knowler WC, Barrett-Connor E, Fowler SE, et al. Reduction in the incidence of type 2 diabetes with lifestyle intervention or metformin. N Engl J Med 2002;346:393–403.
43. Grundy SM, Vega GL, Yuan Z, et al. Effectiveness and tolerability of simvastatin plus fenofibrate for combined hyperlipidemia (the SAFARI trial). Am J Cardiol 2005;95:462–8.
44. The Action to Control Cardiovascular Risk in Diabetes Study Group. N Engl J Med 2008;358:2545–59. Available at http://content.nejm.org/cgi/content/full/358/24/2545. Accessed July 28, 2008.
45. The ACCORD Study Group. Action to Control Cardiovascular Risk in Diabetes (ACCORD) trial: design and methods. Am J Cardiol 2007;99:21i–33i.
46. American Diabetes Association. Standards of medical care in diabetes—2008. Diabetes Care 2008;31(suppl 1):S12–54.
47. Truesdale KP, Stevens J, Cai J. The effect of weight history on glucose and lipids: the Atherosclerosis Risk in Communities Study. Am J Epidemiol 2005;161:1133–43.
48. Ratner R, Goldberg R, Haffner S, et al. Impact of intensive lifestyle and metformin therapy on cardiovascular disease risk factors in the diabetes prevention program. Diabetes Care 2005;28:888–94.
49. Ziccardi P, Nappo F, Giugliano G, et al. Reduction of inflammatory cytokine

concentrations and improvement of endothelial functions in obese women after weight loss over one year. Circulation 2002;105:804–9.
50. Snow V, Barry P, Fitterman N, et al. Pharmacologic and surgical management of obesity in primary care: a clinical practice guideline from the American College of Physicians. Ann Intern Med 2005;142:525–31.
51. Li Z, Maglione M, Tu W, et al. Meta-analysis: pharmacologic treatment of obesity. Ann Intern Med 2005;142:532–46.
52. James W, Astrup A, Finer N, et al. Effect of sibutramine on weight maintenance after weight loss: a randomised trial. Lancet 2000;356:2119–25.
53. De Petrocellis L, Cascio MG, Di Marzo V. The endocannabinoid system: a general view and latest additions. Br J Pharmacol 2004;141:765–74.
54. Pi-Sunyer FX, Aronne LJ, Heshmati HM, et al. Effect of rimonabant, a cannabinoid-1 receptor blocker, on weight and cardiometabolic risk factors in overweight or obese patients: RIO-North America: a randomized controlled trial. JAMA 2006;295:761–75.
55. Scheen AJ, Finer N, Hollander P, et al. Efficacy and tolerability of rimonabant in overweight or obese patients with type 2 diabetes: a randomised controlled study. Lancet 2006;368:1660–72.
56. Van Gaal LF, Rissanen AM, Scheen AJ, et al. Effects of the cannabinoid-1 receptor blocker rimonabant on weight reduction and cardiovascular risk factors in overweight patients: 1-year experience from the RIO-Europe study. Lancet 2005;365:1389–97.
57. Krauss RM, Winston M, Fletcher BJ, et al. Obesity: impact on cardiovascular disease. Circulation 1998;98:1472–6.

5

PHARMACOTHERAPY

Barbara S. Wiggins, Pharm.D., BCPS (AQ Cardiology), CLS, FAHA

Several agents exist for the management of lipid disorders. These include the 3-hydroxy-3-methylglutaryl coenzyme A (HMG-CoA) reductase inhibitors, cholesterol absorption inhibitors, bile acid resins, fibric acid derivatives, nicotinic acid, and omega-3 fatty acids. The choice of agent is dependent on the patient's type of dyslipidemia, cardiovascular risk, comorbidities, as well as the pharmacokinetic profile, cost, and potential for drug interactions. This chapter will provide an overview of the mechanism of action of the various lipid-lowering agents, their pharmacokinetics, common adverse effects, and drug interactions, as well as dosing and administration recommendations.

STATINS

The HMG-CoA reductase inhibitors, commonly referred to as statins, are the most effective and most widely prescribed agents for lowering LDL-C. They have demonstrated clinical benefits across many patient populations and have demonstrated a significant reduction in morbidity and mortality. Therefore, the statins are considered first-line pharmacologic therapy for the management of dyslipidemia in almost every patient with atherosclerotic vascular disease (AVD) and in most patients at risk of developing AVD when dyslipidemia is unable to be adequately managed by TLCs alone. Six statins are currently available: pravastatin, fluvastatin, lovastatin, simvastatin, atorvastatin, and rosuvastatin. Both Figure 1 and Table 1[1-6] summarize the effects of the different statin doses on LDL-C and other lipid parameters.

MECHANISM OF ACTION

All statins share the same mechanism of action. They competitively inhibit the enzyme HMG-CoA reductase, the hepatic enzyme involved in the rate-limiting conversion of HMG-CoA to mevalonate, a precursor of sterol production (Figure 2). Because of the reduced production of mevalonate and its subsequent conversion to cholesterol, compensatory mechanisms increase the synthesis of hepatic LDL receptors. This leads to an increase in LDL and VLDL particle uptake and thus lowers LDL-C, apolipoprotein B, triglyceride, and total cholesterol concentrations. Statins may modestly increase HDL-C concentrations.[7-13]

The primary use of statin agents is to lower LDL-C. Low-density lipoprotein cholesterol is formed from VLDLs and catabolized primarily through the high-affinity LDL receptor. Inhibition of the HMG-CoA reductase enzyme leads to a reduction in cholesterol synthesis in the liver, which then results in an increase in the number of hepatic LDL receptors on the cell surface. This up-regulation leads to enhanced uptake and catabolism of LDL and VLDL particles from the plasma.[1-6]

The statins have several additional "pleiotropic" properties that may play a role in decreasing AVD; these properties are independent of LDL-C lowering. These properties include anti-inflammatory effects; improvement in endothelial function; increased nitric oxide bioavailability; antioxidant effects; stabilization of vulnerable plaques by replacement of the highly lipid core with smooth muscle cells, which leads to a strengthening of the lesion; reduction of plaque progression; and antithrombotic

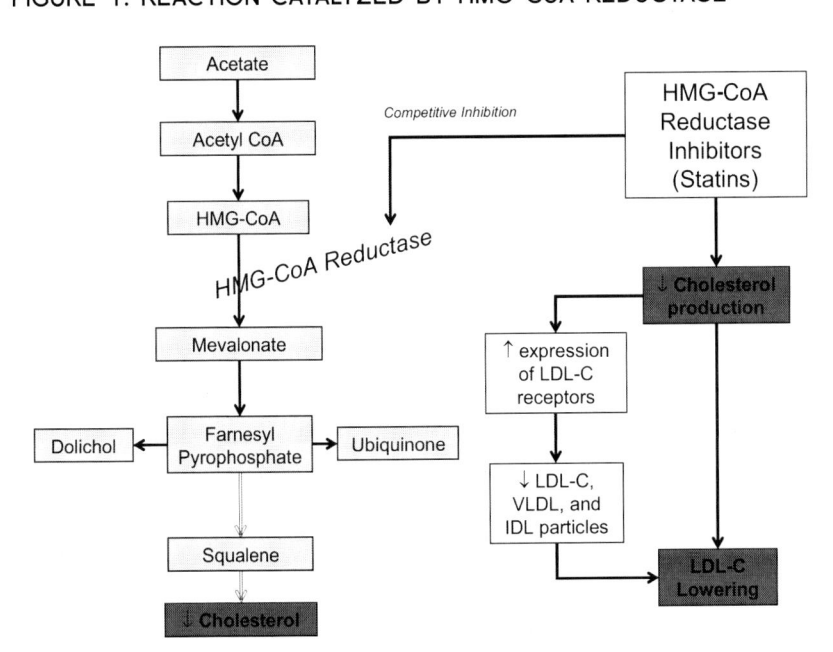

FIGURE 1. REACTION CATALYZED BY HMG-CoA REDUCTASE

HMG-CoA = 3-hydroxy-3-methylglutaryl coenzyme A.

TABLE 1. EFFECT OF STATINS ON LIPID PARAMETERS[1-6]

		TC (%)	LDL-C (%)	HDL-C (%)	TG (%)
Pravastatin	10 mg	−16	−22	+7	−15
	20 mg	−24	−32	+2	−11
	40 mg	−25	−34	+12	−24
	80 mg	−27	−37	+3	−19
Fluvastatin	20 mg	−17	−22	+3	−12
	40 mg	−19	−25	+4	−14
	80 mg[a]	−25	−35	+7	−19
Lovastatin	10 mg	−16	−21	+5	−10
	20 mg	−19	−27	+6	+9
	40 mg	−22	−31	+5	−8
Simvastatin	10 mg	−23	−30	+12	−15
	20 mg	−28	−38	+8	−19
	40 mg	−31	−41	+9	−18
	80 mg	−36	−47	+8	−24
Atorvastatin	10 mg	−29	−39	+6	−19
	20 mg	−33	−43	+9	−26
	40 mg	−37	−50	+6	−29
	80 mg	−45	−60	+5	−37
Rosuvastatin	5 mg	−33	−45	+13	−35
	10 mg	−36	−52	+14	−10
	20 mg	−40	−55	+8	−23
	40 mg	−46	−63	+10	−28

[a]Fluvastatin extended-release preparation.
HDL-C = high-density lipoprotein cholesterol; LDL-C = low-density lipoprotein cholesterol; TC = total cholesterol; TG = triglycerides.

effects.[14] Although all of these so-called pleiotropic effects of statins are considered a "class effect," individual statin agents differ with regard to potency and pharmacokinetic profiles.

Of the available statins, rosuvastatin is the most efficacious in lowering LDL-C. It reduces LDL-C by more than 60% at the maximum dosage of 40 mg once daily (see Figure 1). In comparison, atorvastatin is the next most effective agent, reducing LDL-C by up to 60% at the maximum dosage of 80 mg once daily. The relative efficacy in LDL-C lowering is then followed by simvastatin, lovastatin, pravastatin, and fluvastatin on a milligram per milligram basis.

PHARMACOKINETICS/PHARMACODYNAMICS

The ability of both rosuvastatin and atorvastatin to substantially lower LDL-C is most likely related to their long half-life (see Table 2). This property enables longer inhibition of the HMG-CoA enzyme, therefore resulting in more effective LDL-C lowering than the other statins. This long half-life also enables administration any time of day. This is in contrast to simvastatin, lovastatin, pravastatin, and fluvastatin, which should ideally be administered at bedtime for maximum efficacy. Bedtime administration is

TABLE 2. PHARMACOKINETIC DIFFERENCES AMONG STATINS[1-6]

	Bioavailability (%)	Elimination Half-life (hours)	Metabolism	Prodrug	Solubility
Pravastatin	17	1.5–2	N/A	No	Hydrophilic
Fluvastatin	24	1	2C9	No	Hydrophilic
Lovastatin	< 5	2–3	3A4	Yes	Lipophilic
Simvastatin	< 5	2	3A4	Yes	Lipophilic
Atorvastatin	12	14	3A4	No	Lipophilic
Rosuvastatin	20	20	2C19	No	Hydrophilic

FIGURE 2. CHANGES IN LDL-C WITH DIFFERENT DOSES OF STATIN AGENTS

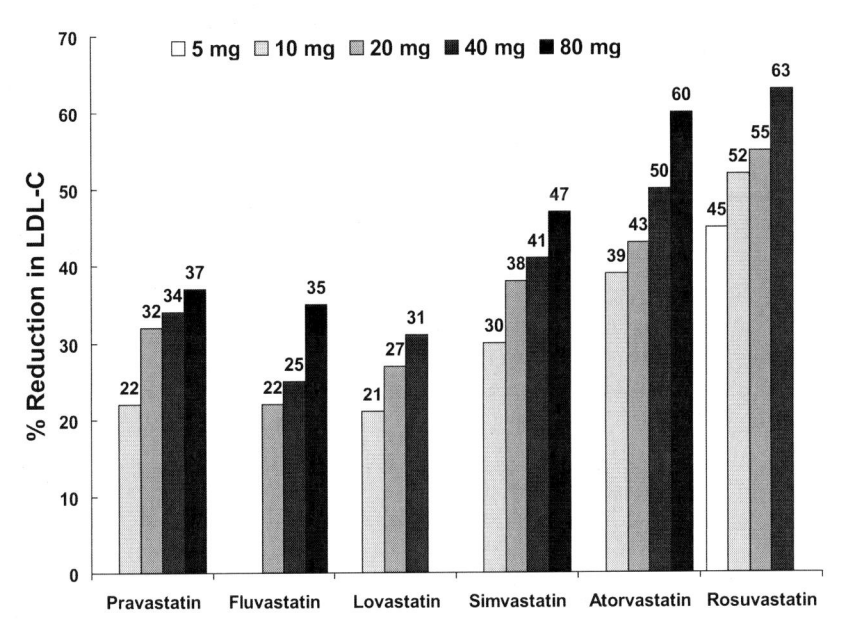

(Reductions are from references 1–6). The fluvastatin 80-mg dose uses the extended-release preparation.
LDL-C = low-density lipoprotein cholesterol.

important for these short half-life statins because cholesterol biosynthesis occurs at its highest rate in the evening hours, thus allowing these agents to have their greatest effect on lowering LDL-C. Based on a few small studies, the difference in LDL-C lowering compared with nighttime administration can be as much as 13% versus if these agents are administered in the morning.[15,16] Moreover, immediate-release lovastatin should be administered with the evening meal to enhance the bioavailability of the drug.[3] With extended-release formulations of lovastatin and fluvastatin, bedtime administration is less important.

TABLE 3. DOSING OF STATIN AGENTS IN CKD[1-6]

	GFR (mL/minute)		
Drug	30–59	15–29	< 15 or HD
Atorvastatin	10–80 mg	10–80 mg	10–80 mg
Fluvastatin	1–80 mg	10–40 mg	10–40 mg
Lovastatin	20–80 mg	10–40 mg	10–40 mg
Pravastatin	20–40 mg	20–40 mg	20–40 mg
Rosuvastatin	5–40 mg	5–10 mg	—[a]
Simvastatin	20–80 mg	10–40 mg	10–40 mg

[a]No data available – cannot recommend.
GFR = estimated glomerular filtration rate; HD = hemodialysis.

Oral absorption of the statins varies considerably.[1-6] The bioavailability ranges from about 30% for lovastatin to 98% for fluvastatin. A comparison of the various pharmacokinetic parameters of these agents is listed in Table 2.

Hydrophilic agents (i.e., pravastatin, fluvastatin, and rosuvastatin) dissolve more readily in the gastrointestinal (GI) tract; however, absorption across the intestinal mucosa is difficult. These agents also tend to have lower rates of presystemic elimination. All of the statins are eliminated primarily by the liver, with substantial biliary excretion. However, significant renal dysfunction (creatinine clearance [CrCl] less than 30 mL/minute) can lead to increased risk of statin-associated myopathy. Dosing recommendations for these agents in patients with renal insufficiency are given in Table 3.

Atorvastatin, lovastatin, and simvastatin are substrates for the cytochrome P450 (CYP) 3A4 isoenzyme, which makes them more prone to drug interactions than the other statins. Fluvastatin is primarily a substrate of the CYP2C9 isoenzyme and, to a lesser extent, CYP2C8 and CYP3A4. Atorvastatin, lovastatin, and simvastatin also have active metabolites that contribute to their lipid-lowering effects. Fluvastatin has an active metabolite, but it does not enter the circulation and thus does not contribute to the drug's efficacy. Pravastatin and rosuvastatin are the only statins not extensively metabolized by the CYP enzyme system. Pravastatin is broken down through sulfation and phase 2 reactions involving conjugation with other agents. About 10% of rosuvastatin is metabolized, with the primary isoenzymes involved being CYP2C9 and CYP2C19. The primary metabolite *N*-desmethyl rosuvastatin has one-sixth to one-half the HMG-CoA reductase activity of rosuvastatin.[5]

ADVERSE EFFECTS

Statins are generally well tolerated, with the most commonly reported adverse effects being GI symptoms, headache, and skin rashes. Although not as common, adverse effects that are of most concern to health care professionals are hepatotoxicity, elevated hepatic transaminases, myopathy, and proteinuria.

Hepatotoxicity

Rare cases of acute liver failure have been reported with all of the statins. The incidence of liver failure in statin-treated patients is about 1 per 1 million patients treated. Of interest, this is the same as the background rate of acute liver failure (in patients not taking statins). Elevated hepatic transaminase concentrations (more than 3 times the upper limit of normal [ULN]) are dose-dependent and usually reversible with dosage reduction or discontinuation. However, in 70% of cases, the transaminase abnormality will resolve with continued statin use.[17] Switching to another statin or a rechallenge with the same drug usually does not result in a recurrence of the hepatic transaminase elevation, and no cases of cross-reactivity have been published. If a hepatic transaminase elevation is detected, the test should be repeated. Only persistent elevations (at least twice more) should result in changing therapy.

In the setting of preexisting liver dysfunction, the statin agents are only contraindicated in patients with decompensated cirrhosis or in patients with persistent unexplained hepatic transaminase elevations. Some liver experts believe that the statins may be used safely in patients with nonalcoholic fatty liver disease and nonalcoholic steatohepatitis.[18] In addition, no conclusive data have shown that statins exacerbate liver disease or elevated hepatic transaminase concentrations in patients with hepatitis B or C.

Muscle-Related Toxicity

All statins have also been associated with muscle-related adverse effects. The three types of muscle-related adverse effects are (1) myalgia (muscle pain, soreness, or weakness), (2) myopathy (myalgia *plus* an elevation in serum creatine kinase (CK) of more than 10 times ULN), and (3) rhabdomyolysis (CK of 10 or more times ULN with an SCr elevation or requiring treatment for rhabdomyolysis).[17] Rhabdomyolysis is considered rare but is potentially life threatening because it can result in acute renal failure secondary to myoglobinuria, cardiac arrest, or arrhythmias caused by hyperkalemia and hypocalcemia. Recommendations from the National Lipid Association regarding muscle and statin safety are listed in Table 4.[17] Common risk factors for statin-related muscle adverse effects include advanced age, frailty, CKD, interacting drugs, and hypothyroidism.

Kidney Toxicity

Statins reportedly cause proteinuria. This issue was first raised with reports from preclinical data involving rosuvastatin in which an incidence of 12–15% was observed with the 80-mg dose.[19] However, this is twice the maximum dose currently approved by the FDA.[19] It now appears that the proteinuria observed with statins is no more than a normal physiologic consequence due to the inhibition of the HMG-CoA reductase enzyme. When inhibition of the HMG-CoA reductase enzyme occurs, there is a reduction in available mevalonate, which is a key component of renal tubular protein reabsorption. When mevalonate decreases, low-molecular-weight proteins are secreted into the urine. Secretion of these small proteins is not indicative of kidney dysfunction or injury; when the mevalonate concentrations are restored, these small

TABLE 4. NATIONAL LIPID ASSOCIATION 2006: RECOMMENDATIONS TO HEALTH PROFESSIONALS REGARDING THE MUSCLE AND STATIN SAFETY

1. Whenever muscle symptoms or an increased CK concentration is encountered in a patient receiving statin therapy, health professionals should attempt to rule out other etiologies, because these are most likely to explain the findings. Other common etiologies include increased physical activity, trauma, falls, accidents, seizure, shaking chills, hypothyroidism, infections, carbon monoxide poisoning, polymyositis, dermatomyositis, alcohol abuse, and drug abuse (cocaine, amphetamines, heroin, or phencyclidine).
2. Obtaining a pretreatment, baseline CK concentration may be considered in patients who are at high risk of experiencing a muscle toxicity (e.g., older individuals; when combining a statin with an agent known to increase myotoxicity), but this is not routinely necessary in other patients.
3. It is not necessary to measure CK concentrations in asymptomatic patients during statin therapy, because marked, clinically important CK elevations are rare and are usually related to physical exertion or other causes.
4. Patients receiving statin therapy should be counseled about the increased risk of muscle complaints, particularly if the initiation of vigorous, sustained endurance exercise or a surgical operation is being contemplated; they should be advised to report such muscle symptoms to a health professional.
5. CK measurements should be obtained in symptomatic patients to help gauge the severity of muscle damage and facilitate a decision of whether to continue therapy or alter doses.
6. In patients who develop intolerable muscle symptoms with or without a CK elevation and in whom other etiologies have been ruled out, the statin should be discontinued. Once asymptomatic, the same or different statin at the same or lower dose can be restarted to test the reproducibility of symptoms. Recurrence of symptoms with multiple statins and doses requires initiation of other lipid-altering therapy.
7. In patients who develop tolerable muscle complaints or are asymptomatic with a CK < 10 the ULN, statin therapy may be continued at the same or reduced doses and symptoms may be used as the clinical guide to stop or continue therapy.
8. In patients who develop rhabdomyolysis (a CK > 10,000 IU/L or a CK > 10 times the ULN with an elevation in SCr or requiring intravenous hydration therapy), statin therapy should be stopped. Intravenous hydration therapy in a hospital setting should be instituted if indicated for patients experiencing rhabdomyolysis. Once recovered, the risk vs. benefit of statin therapy should be carefully reconsidered.

Adapted from McKenney JM, Davidson MH, Jacobson TA, Guyton JR. Final conclusions and recommendations of the National Lipid Association statin safety assessment task force. Am J Cardiol 2006;97:S89–S94.

proteins are then absorbed across the tubules.[20,21] Based on current data, there appears to be no relationship between nephrotoxicity and statins.

DRUG INTERACTIONS

The drug interaction profile for individual statin agents is an important factor to consider when selecting pharmacotherapy. Statins not significantly metabolized by CYP isoenzymes (e.g., pravastatin, rosuvastatin) or those with a high bioavailability (fluvastatin) are the least prone to pharmacokinetic drug interactions. Lovastatin, simvastatin, fluvastatin, and, to a lesser extent, atorvastatin are metabolized by the CYP isoenzyme system. Lovastatin, simvastatin, and atorvastatin are all substrates of isoenzyme CYP3A4. They can interact with other drugs that inhibit this isoenzyme. They may interact with drugs that also require metabolism through the CYP3A4, although these types of interactions are not well documented in the literature and are more theoretical. A variety of agents, including protease inhibitors, calcium channel blockers, azole antifungal agents such as ketoconazole and itraconazole, cyclosporine, erythromycin, and large quantities of grapefruit juice (more than 1 quart/day), are inhibitors of CYP3A4. They can inhibit the metabolism of atorvastatin, lovastatin, and simvastatin, resulting in increased serum concentrations and an increased risk of drug toxicity. Although not a substrate of CYP3A4, rosuvastatin interacts with cyclosporine and gemfibrozil, resulting in significant increases in serum concentrations of rosuvastatin.[5]

Using statins in combination with gemfibrozil can also result in clinically significant drug interactions. Gemfibrozil inhibits glucuronidation, an elimination pathway of atorvastatin, lovastatin, simvastatin, pravastatin, and rosuvastatin. This leads to elevated concentrations of these statins, thus increasing the risk of toxicity. In contrast, fenofibrate appears to have a minimal effect on the metabolism of these statins and is the safer alternative in patients who require combination therapy.[22,23]

Pravastatin is not significantly metabolized by the CYP isoenzyme system, and it is devoid of many of the drug interactions observed with the other statins. Fluvastatin is a substrate of CYP2C9, which explains why its drug interaction profile is also different. Rosuvastatin is a substrate of CYP2C9 and CYP2C19 and undergoes minimal hepatic metabolism. Available evidence suggests that both rosuvastatin and pravastatin have a low potential for interacting with drugs that inhibit the CYP metabolic pathways. A review of select drug interactions involving the statin agents is listed in Table 5.

DOSING AND ADMINISTRATION

Statin therapy should be used as adjunctive therapy to lifestyle modifications (diet and exercise). Immediate-release lovastatin should be administered with the evening meal to enhance absorption. However, extended-release lovastatin has maximum efficacy when given at bedtime. Pravastatin, simvastatin, fluvastatin, atorvastatin, and rosuvastatin may be administered without regard to food. Because of their relatively short half-lives, the statins have traditionally been administered at night, so their peak effect corresponds to the height of endogenous cholesterol synthesis. Atorvastatin and rosuvastatin are the exceptions; they may be dosed at any time of day because of their longer half-lives.

CHOLESTEROL ABSORPTION INHIBITORS

Ezetimibe (Zetia) is the only agent currently available in this class. It may be used alone or in combination with an HMG-CoA reductase inhibitor or fenofibrate together with diet for the management of dyslipidemia, specifically to lower LDL-C.

MECHANISM OF ACTION

Ezetimibe selectively inhibits the intestinal absorption of cholesterol by localizing at the brush border of the small intestine, which then leads to reduced delivery of cholesterol to the liver.[24] A reduction in the delivery of cholesterol to the liver results

TABLE 5. SELECT DRUG INTERACTIONS WITH THE STATINS[a]

	Lovastatin	Pravastatin	Simvastatin	Fluvastatin	Atorvastatin	Rosuvastatin
Cholestyramine		↓ Pravastatin AUC (40–50%) Administer pravastatin 1 hour before or 4 hours after cholestyramine.		↓ Fluvastatin AUC (51–89%) when given within 4 hours of cholestyramine ↓ Cmax (50–90%)		
Colestipol		↓ Pravastatin AUC (40% to 50%) Administer pravastatin 1 hour before or 4 hours after cholestyramine.			↓ Atorvastatin plasma concentrations (25%) No effect on LDL reduction	
Cyclosporine	↑ Risk of myopathy and rhabdomyolysis ↑ Cyclosporine AUC (20-fold)	↑ Pravastatin AUC (20-fold) ↑ Pravastatin Cmax (7-fold) Daily doses of pravastatin ≥ 10 mg should be administered with caution.	↑ Simvastatin AUC (3-fold) Daily dose of simvastatin should not exceed 10 mg.	↑ Fluvastatin AUC 1.9 fold ↑ C_{max} (1.3-fold)	↑ Risk of myopathy and rhabdomyolysis	Daily dosage of rosuvastatin should not exceed 5 mg.
Diltiazem	↑ Lovastatin AUC (up to 4-fold) ↑ Lovastatin C_{max} (50%) ↑ Risk of myopathy	No effects noted.	↑ Simvastatin plasma concentrations ↑ Risk of myopathy	No effect on steady-state plasma concentrations		
Erythromycin	↑ Lovastatin plasma concentrations ↑ Risk of myopathy and rhabdomyolysis Combination should be avoided.	No effects noted.	↑ Simvastatin plasma concentrations ↑ Risk of myopathy and rhabdomyolysis Combination should be avoided.		↑ Atorvastatin plasma concentrations (40%) ↑ Atorvastatin AUC (33%) ↑ Risk of myopathy and rhabdomyolysis	No effects noted.

	Lovastatin	Pravastatin	Simvastatin	Fluvastatin	Atorvastatin	Rosuvastatin
Gemfibrozil	↑ Risk of myopathy and rhabdomyolysis Daily dose of lovastatin should not exceed 20 mg.	↓ Pravastatin urinary excretion and protein binding Combination should be avoided.	↑ Risk of myopathy and rhabdomyolysis Daily dose of simvastatin should not exceed 10 mg.	No effects noted.	↑ Risk of myopathy and rhabdomyolysis Combination should be avoided.	Daily dosage of rosuvastatin should not exceed 10 mg.
Grapefruit juice (> 1 quart/day)[b]	↑ Lovastatin Cmax (12-fold) ↑ Lovastatin AUC (1.3- to 15-fold) ↑ Risk of myopathy and rhabdomyolysis	No effects noted.	↑ Simvastatin Cmax (9-fold) ↑ Simvastatin AUC (16-fold)		↑ Atorvastatin AUC (3-fold) ↑ t_{max} (1–4 hours) ↑ $t_{1/2}$ (4–8 hours)	
Itraconazole, ketoconazole, voriconazole	↑ Lovastatin plasma concentrations (10- to 30-fold) ↑ Risk of myopathy Combination should be avoided.	↑ In pravastatin AUC (factor of 1.7) and Cmax (factor of 2.5)	↑ Simvastatin plasma concentration ↑ risk of rhabdomyolysis Combination should be avoided.	No significant effects noted.	↑ Atorvastatin AUC and C_{max} (3- to 4-fold) and C_{max} (2-fold)	Modest interaction; not clinically significant
Protease inhibitors	↑ Lovastatin plasma concentrations Combination should be avoided.		↑ Simvastatin plasma concentrations Combination should be avoided.		↑ Atorvastatin plasma concentrations	
Verapamil	↑ Risk of myopathy and rhabdomyolysis Daily doses of lovastatin should not exceed 40 mg.		↑ Risk of myopathy and rhabdomyolysis Daily dose of simvastatin should not exceed 20 mg.			

[a] Blank boxes indicate that no published drug interaction data are available.
[b] Grapefruit juice can be consumed while on statin therapy; if muscle aches develop, then a reduction in quantity or discontinuation of grapefruit juice should be considered.
AUC = area under the curve; C_{max} = maximum plasma concentration; INR = international normalized ratio; PTT = partial thromboplastin time; t_{max} = time of occurrence for maximum (peak) drug concentration; $t_{1/2}$ = half-life.

Adapted from Lipitor [package insert]. New York: Pfizer; 2006. Fluvastatin sodium (Lescol XL) [package insert]. East Hanover, NJ: Novartis Pharmaceutical; 2006. Lovastatin (Mevacor) [package insert]. Whitehouse Station, NJ: Merck & Co.; 2005. Pravastatin (Pravachol) [package insert]. Princeton, NJ: Bristol-Myers Squibb; 2002. Rosuvastatin (Crestor) [package insert]. Wilmington, DE: AstraZeneca Pharmaceuticals; 2005. Simvastatin (Zocor) [package insert]. Whitehouse Station, NJ: Merck & Co.; 2005.

TABLE 6. EFFECT OF EZETIMIBE ON LIPID PARAMETERS

	TC (%)	LDL-C (%)	TG (%)	HDL-C (%)
Ezetimibe[a] alone	−12	−18	−7	+1
Ezetimibe[a,b] + HMG-CoA	−17	−25	−14	+3
Ezetimibe + fenofibrate 160 mg	−22	−20	−44	+19

[a]Mean percent change from baseline.
[b]Represents mean of all statins.
TC = total cholesterol; LDL-C = low-density lipoprotein cholesterol; HDL-C = high-density lipoprotein cholesterol; TG = triglycerides.
Source: Ezetimibe (Zetia) [package insert]. North Wales, PA: Merck/Schering-Plough Pharmaceuticals; 2006.

in reduced hepatic cholesterol stores and increased clearance from the blood through up-regulation of the LDL receptor. Monotherapy with ezetimibe can reduce LDL-C by about 18%. When combined with fenofibrate, LDL-C can be reduced by up to 20% and by an average of 25% when combined with a statin.[24] The efficacy of ezetimibe is summarized in Table 6.

PHARMACOKINETICS/PHARMACODYNAMICS

Ezetimibe is a prodrug that is rapidly conjugated to an active phenolic glucuronide (ezetimibe-glucuronide).[24] The drug is primarily metabolized in the small intestine by means of glucuronide conjugation with subsequent renal and biliary excretion. The elimination half-life for both ezetimibe and ezetimibe-glucuronide is about 22 hours. Food does not alter the absorption, and it can be administered at any time of day without regard to meals. Dose adjustments are not necessary for patients with renal or with mild hepatic insufficiency.

ADVERSE REACTIONS

Ezetimibe is well tolerated with minimal adverse effects compared with placebo. The adverse effects most often reported include fatigue, abdominal pain, diarrhea, arthralgia, and back pain. Compared with ezetimibe monotherapy, there is a small increase in the incidence of consecutive hepatic transaminase elevations when ezetimibe is added to a statin, even when the statin dose is unchanged. Therefore, serum hepatic transaminases should be measured before and then 6 weeks after adding ezetimibe to a statin. In addition, the use of ezetimibe alone or in combination with a statin should be avoided in patients with active liver disease or persistent elevations in serum hepatic transaminases that cannot otherwise be explained.[24]

DRUG INTERACTIONS

Ezetimibe has been evaluated in combination with several other drugs to assess potential drug interactions. When combined with digoxin, oral contraceptives, cimetidine, antacids, statins, glipizide, or fenofibrate, there appears to be minimal impact on either drug in terms of an alteration in metabolism and increased bioavailability that requires any additional intervention.[24] Initial reports with warfarin and ezetimibe did not suggest an interaction. However, postmarketing reports indicate there are some elevations in the international normalized ratio (INR) after ezetimibe is added to warfarin. Although the mechanism of this interaction is not understood, additional monitoring may be warranted. The use of cyclosporine in combination with ezetimibe results in increased exposure of both drugs. Although no additional monitoring for ezetimibe is needed, cyclosporine serum concentrations should be followed closely when using this combination, and an initial dose of 5 mg of ezetimibe is recommended.

Using cholestyramine and colestipol concomitantly with ezetimibe results in a significant reduction in the area under the curve of ezetimibe by as much as 80%. Therefore, when the combination of these two agents is needed, the administration of ezetimibe should occur at least 2 hours before or 4 hours after the administration of cholestyramine or colestipol.[24] The combined use of colesevelam with ezetimibe should not result in this type of drug interaction.

BILE ACID SEQUESTRANTS

Three bile acid sequestrants (BASs) are available for use: cholestyramine, colestipol, and colesevelam. At optimal doses, these agents can reduce LDL-C by 15–30%. However, cholestyramine and colestipol are often underused in clinical practice because of palatability and GI intolerance. Colesevelam is better tolerated and is not subject to the classic binding interactions seen with the other two agents; it is generally the preferred BAS agent. All of these agents may be used alone or in combination with other cholesterol-lowering agents. The efficacy of the BASs is summarized in Table 7.[25-27]

TABLE 7. EFFECT OF BAS AGENTS ON CHOLESTEROL PARAMETERS

	TC (%)	LDL-C (%)	TG (%)	HDL-C (%)
BAS alone	−8 to −27	−12 to −37	0 to +12	+3 to +10
BAS + statin	−24 to −40	−45 to −56	−9 to +19	−13 to +18
BAS + fibrate	−22 to −28	−20 to −35	−15 to −44	0 to +11

HDL-C = high-density lipoprotein cholesterol; LDL-C = low-density lipoprotein cholesterol; TC = total cholesterol; TG = triglycerides.

Sources: Cholestyramine (Questran) [package insert]. Princeton, NJ: Bristol-Myers Squibb; 1997.
Colestipol (Colestid) [package insert]. Kalamazoo, MI: Pharmacia & Upjohn; 1999.
Colesevelam hydrochloride (Welchol) [package insert]. Parsippany, NJ: Daiichi Sankyo; 2006.

MECHANISM OF ACTION

Cholesterol is the sole precursor of bile acids. With normal digestion, bile acids are secreted into the intestines through the bile. Bile acids are responsible for emulsifying fat and lipid particles in food, thus facilitating absorption. Most bile acids that are secreted into the GI tract are reabsorbed in the intestines and returned to the liver by enterohepatic circulation. The BASs work by exchanging chloride ions for bile acids and other anions within the intestinal lumen. When bile acids are bound by these agents, their enterohepatic recirculation is interrupted, and their fecal excretion is increased. The fecal loss of the bile acids leads to increased conversion of cholesterol in the hepatocyte to bile acids by the action of 7α-hydroxylase, resulting in an increase in the number of LDL receptors, increased hepatic update of systemic LDL particles, and lowered LDL-C.

PHARMACOKINETICS/PHARMACODYNAMICS

Colestipol and cholestyramine are nonabsorbable hydrophilic basic anion exchange resins that are insoluble in water. Colesevelam is also a nonabsorbable polymeric compound that binds to bile acids more strongly and specifically than the other two agents. However, the drug is also hydrophilic and insoluble in water.

ADVERSE EFFECTS

The most common and problematic adverse effect of these agents is constipation. The incidence increases with the dose and age of the patient, which can limit their use in certain patients. These agents are contraindicated in patients with complete biliary obstruction because bile is not secreted into the intestines. Additional GI adverse effects include abdominal pain, flatulence, nausea, vomiting, dyspepsia, and steatorrhea.[25-27] The incidence of GI effects is less with colesevelam than with the other agents. This is most likely because the drug is more effective at lower doses. Patients taking these agents may theoretically be at increased risk of bleeding caused by hypoprothrombinemia from vitamin K deficiency. If this adverse effect occurs, patients can be treated by the administration of intravenous vitamin K_1, and prevention can be undertaken by administering daily vitamin K_1 orally.

DRUG INTERACTIONS

Cholestyramine and colestipol may delay or reduce the absorption of many drugs administered concomitantly.[25,26] Therefore, other drugs should be taken at least 1 hour before or 4 hours after taking one of these two agents; this includes other cholesterol-lowering agents. Colesevelam does not have these interactions. It can be taken concurrently with other drugs, including fenofibrate and HMG-CoA reductase inhibitors, without any impact on efficacy.[27] The concurrent use of colesevelam with levothyroxine has not been well studied. Postmarketing surveillance data have demonstrated that some patients on concurrent colesevelam with levothyroxine therapy have an increase in thyroid-stimulating hormone. Therefore, patients receiving both of these agents should have their thyroid-stimulating hormone values closely monitored.

DOSING AND ADMINISTRATION

The recommended initial dose of colesevelam is 3 tablets twice daily or 6 tablets once daily taken with a liquid and a meal.[27] Cholestyramine oral suspensions should be initiated with one packet (4 g) once or twice daily, with the dose gradually increased to a maintenance dose of 8–16 g divided into two doses.[25] Each packet should be mixed with 2–3 oz of water or a beverage of choice before consumption. In addition to liquids, the drug may be mixed with applesauce, crushed pineapple, soups, or pulpy fruits. The drug may be taken with or without food but should not be taken together with other drugs. If other drugs are required, they should be taken 1 hour before or 4 hours after cholestyramine. Colestipol is available as a tablet and as an oral suspension. The oral suspension dose should be initiated at a dosage of one packet daily and increased by one dose per day at 1- to 2-month intervals.[26] The usual adult dosage is one to six packets once or twice daily. Like cholestyramine, colestipol should always be mixed with water or another beverage and should be separate from other drugs as previously described for cholestyramine. Colestipol tablets should be taken at a dose of 2 g once or twice daily with a maintenance dose ranging from 2 to 16 g. The tablets should be taken one at a time and should not be crushed or chewed. The effect of the BASs on lipoprotein parameters is listed in Table 8. The BASs are not systemically absorbed, and no dose adjustments are required for patients with renal insufficiency. Given that BASs can increase triglycerides, these agents should only be initiated if the patient's triglyceride concentration is less than 200 mg/dL and should be avoided if the triglyceride concentration is higher than 400 mg/dL.

FIBRIC ACID DERIVATIVES

Fenofibrate and gemfibrozil are the only two fibric acid derivatives used in clinical practice. Fenofibrate should be the preferred agent because it has fewer drug interactions. The primary role of these agents is to treat hypertriglyceridemia.

TABLE 8. DOSE-DEPENDENT EFFECTS OF BAS ON LDL-C CONCENTRATIONS

	Dose (g)	LDL-C Reduction (%)
Colestipol	5	−12
	10	−20
	15	−24
Cholestyramine	4–8	−9
	8–12	−13
	12–16	−17
	16–20	−21
	20–24	−28
Colesevelam	3.8	−15
	4.5	−18

BAS = bile acid sequestrant; LDL-C = low-density lipoprotein cholesterol.

MECHANISM OF ACTION

The fibric acid derivatives activate peroxisome proliferator activated receptor α (PPAR-α). Activation of this receptor leads to an increase in lipolysis and elimination of triglyceride-rich particles through the activation of lipoprotein lipase and reduction in the production of apoprotein C-III. Serum triglyceride reductions lead to an alteration in the size and composition of LDL-C, as LDL-C particle size is transformed from small and dense to large and buoyant. The small, dense LDL-C particle is the most atherogenic because of its susceptibility to oxidation.

PHARMACOKINETICS/PHARMACODYNAMICS

Fenofibrate is well absorbed from the GI tract with exact bioavailability unknown.[28] It is primarily metabolized by conjugation with peak plasma concentrations occurring 6–8 hours after oral administration. Fenofibrate is highly bound to plasma proteins, with an elimination half-life of 20 hours. The drug is excreted primarily in the urine in the form of metabolites.

Gemfibrozil is completely absorbed after oral administration and reaches peak plasma concentrations within 1–2 hours.[29] The rate and extent of absorption of gemfibrozil is best when taken 30 minutes before meals. Taking the drug after meals can result in a 14–44% reduction in the area under the curve. Gemfibrozil is highly bound to plasma proteins, primarily albumin, with an elimination half-life of 1.5 hours. The drug undergoes extensive hepatic metabolism by oxidation of the ring methyl group to form hydroxymethyl and carboxyl metabolites. About 70% of gemfibrozil is renally eliminated predominantly as the glucuronide conjugate.[29]

ADVERSE EFFECTS

The most common adverse effects associated with fibric acid derivatives include nausea, vomiting, dyspepsia, diarrhea, abdominal pain, flatulence, and constipation. Fibric acid derivatives have also been associated with abnormalities in liver function tests including bilirubin and alkaline phosphatase.[28,29] However, when elevations are seen with these agents, they will often return to baseline after drug discontinuation. Myalgia has also been reported with these agents. The incidence of myalgia and more serious muscle-related adverse effects (myopathy and rhabdomyolysis) is increased when the fibrates are used in combination with a statin, warranting careful monitoring when the combination is used.

A rise in SCr concentrations after initiating a fibric acid derivative, primarily with fenofibrate, has also been documented.[30-34] No resultant decrease in the GFR appears to be present, suggesting that the increase in SCr does not represent serious kidney toxicity resulting in intrinsic kidney dysfunction.[30,35,36] The exact mechanism by which this occurs is not fully understood, but it may be due to their effect on PPAR-α.[37] Specific dosing recommendations for the use of these agents in patients with impaired renal function are discussed below.

DRUG INTERACTIONS

Drug interactions associated with fibric acid derivative are fairly well known; several of these drug interactions can be managed by careful monitoring. The most clinically significant interactions occur with statins, warfarin, repaglinide, cholestyramine, and colestipol. The combination of statins and gemfibrozil increases systemic concentrations of the statin and raises the risk of developing myopathy and rhabdomyolysis. As previously mentioned, data suggest that there is less risk of a significant drug interaction with fenofibrate and, subsequently, a lower risk of myopathy and rhabdomyolysis. Gemfibrozil and fenofibrate should not be administered with lovastatin in doses higher than 20 mg or simvastatin in doses higher than 10 mg. Gemfibrozil should not be administered with rosuvastatin in doses higher than 10 mg. However, there are no data demonstrating that adding a fibrate to a statin will reduce CHD risk. The co-administration of gemfibrozil with ezetimibe may cause increased cholesterol excretion into the bile, thereby increasing the risk of cholelithiasis.[29] This does not appear to be a concern with fenofibrate in combination with ezetimibe.

Concurrent use of warfarin with a fibric acid derivative results in an increased risk of bleeding. The exact mechanism of this interaction is not well established but is believed to result from the displacement of warfarin from protein binding. Therefore, INR values should be carefully followed in patients on both of these agents. An empiric reduction in the dose of warfarin may be justified before initiating a fibric acid derivative to minimize the risk of bleeding complications.

Repaglinide should be used cautiously in combination with gemfibrozil because the combination may result in enhanced and prolonged hypoglycemic effects.[29] If combination therapy is deemed necessary, blood glucose should be closely monitored and the dose of repaglinide decreased as needed.

DOSING AND ADMINISTRATION

The initial dose of fenofibrate in patients with primary hypercholesterolemia or mixed hyperlipidemia is usually the highest dose available. Because there are many different formulations of fenofibrate available (e.g., TriCor, Lofibra, Antara), this dose ranges from 130 to 200 mg. All products are dosed once daily. For the treatment of hypertriglyceridemia, the initial dose may be at the lower end of the dosing range. If the initial dose chosen is lower than the maximum dose, it should be titrated upward at 4- to 8-week intervals, based on patient response, up to the maximum dose. In patients with renal insufficiency or in elderly patients, the initial dose of fenofibrate should be at one-third of the maximum daily dosage and should only be titrated after careful monitoring of renal function and lipid concentrations at that dose. The National Kidney Foundation and The National Lipid Association recommend avoiding the use of fenofibrate when the CrCl is less than 15 mL/minute.[38,39]

For gemfibrozil, the recommended adult dose is 600 mg twice daily administered at least 30 minutes before the morning and evening meals. In patients with severe hepatic insufficiency, severe CKD, preexisting gallbladder disease, and primary biliary cirrhosis, gemfibrozil is contraindicated. The National Kidney Foundation recommends gemfibrozil as the preferred fibrate to be used in patients with chronic kidney failure or in patients with prior kidney transplantation.[38] However, the National Lipid Association Safety Task Force recommends a 50% reduction in dose if the estimated GFR is 15–59 mL/minute/1.73 m^2 and an avoidance of use if the estimated GFR is less than 15 mL/minute/1.73 m^2.[39]

TABLE 9. NATIONAL LIPID ASSOCIATION RECOMMENDATIONS FOR THE SAFE PRESCRIBING OF FIBRATE THERAPY

1. Obtain a baseline SCr before the initiation of therapy.
2. In the presence of impaired renal function, gemfibrozil is preferred.
3. Consider discontinuation or dose reduction of fibrate therapy if in the setting of increased SCr if other etiologies have been ruled out.
4. Fenofibrate is the preferred agent to use when combination with a statin is warranted. If gemfibrozil is used, adhere to manufacturer maximum dosing recommendations.
5. Avoid the use of gemfibrozil in patients with renal dysfunction also taking statins that undergo a high degree of renal elimination.
6. Obtain a baseline CK before adding a fibrate to statin therapy and in those patients at high risk.
7. Inform patient to report any potential signs of muscle toxicity.
8. Closer monitoring of the INR is recommended because of the possible potentiation of anticoagulation.
9. Be aware of the risk of cholelithiasis, and if gallstones are present, consider drug discontinuation.
10. Be aware that fibrates can increase homocysteine concentrations and that routine monitoring is not necessary.

Adapted from Davidson MH, Armani A, McKenney JM, et al. Safety considerations with fibrate therapy. Am J Cardiol 2007;99(suppl):3C–18C.

Fenofibrate should be administered at least 1 hour before or 4–6 hours after cholestyramine to avoid reduced absorption of fenofibrate. This interaction does not appear to occur with gemfibrozil. Recommendations on the safe use of fibrate therapy in patients with hypertriglyceridemia are listed in Table 9.

NICOTINIC ACID

Nicotinic acid (niacin) is a water-soluble B vitamin that plays an essential role in energy metabolism and DNA repair. While niacin is an essential nutrition source, at therapeutic dosages (1–2 g/day), niacin is an antihyperlipidemic agent. Niacin is available in three dosage forms: immediate release, slow release, and extended release. The extended-release tablet (i.e., Niaspan) is the only prescription niacin product and is the most effective agent available for raising HDL-C.

MECHANISM OF ACTION

The exact mechanism by which niacin affects cholesterol is not fully understood. However, the process may involve partial inhibition of the release of free fatty acids from adipose tissue. This results in increased lipoprotein lipase activity that may then increase the rate of chylomicron triglyceride removal from the plasma. Niacin also decreases the rate of hepatic synthesis of both VLDL-C and LDL-C. Niacin increases HDL-C by lowering plasma triglycerides and decreasing the rate of HDL catabolism through reverse cholesterol transport.[40] The increase in HDL-C is associated with an increase in apolipoprotein A-1 and a shift in the distribution of HDL subfractions,

TABLE 10. DOSE-DEPENDENT EFFECTS OF EXTENDED-RELEASE NIACIN (NIASPAN) ON LIPID PARAMETERS

	TC (%)	LDL-C (%)	TG (%)	HDL-C (%)
Niaspan 500 mg	−2	−3	−6	+10
Niaspan 1000 mg	−5	−9	−5	+15
Niaspan 1500 mg	−11	−14	−28	+22
Niaspan 2000 mg	−12	−17	−35	+26

HDL-C = high-density lipoprotein cholesterol; LDL-C = low-density lipoprotein cholesterol; TC = total cholesterol; TG = triglycerides.
Source: Niacin (Niaspan) [package insert]. Cranbury, NJ: Kos Pharmaceuticals; 2005.

resulting in an increase in the HDL 2:HDL 3 ratio. An increase in HDL 2 represents an increase in the large buoyant particle size that has been most associated with lower risk of AVD. Niacin also reduces apolipoprotein B-100 concentrations, which is the major component of VLDL, LDL fractions, and Lp(a). The efficacy of extended release is summarized in Table 10.

PHARMACOKINETICS/PHARMACODYNAMICS

Niacin is absorbed rapidly and extensively, with a bioavailability of 60–76%, and the lipid-lowering effects are dose-dependent.[40] The drug undergoes rapid and extensive first-pass metabolism. Both niacin and its metabolites are eliminated in the urine.

ADVERSE EFFECTS

All of the niacin preparations have adverse effects that may complicate adherence. The most common adverse effect is flushing that occurs more commonly with the immediate-release formulation and the least with the extended-release product. The flushing is often described as a prickly heat feeling and is not associated with diaphoresis and rarely hypotension.[41] The onset of flushing is usually quicker and shorter in duration with the immediate-release preparation than the extended release. This adverse effect occurs because of the release of prostaglandin D_2 from the skin.[42] However, with continued use, tachyphylaxis develops, and flushing diminishes in frequency and intensity over time with continued administration.[43] Flushing may also be worsened by hot spicy foods, hot beverages, or hot baths or showers.[41] Some over-the-counter (OTC) products claim to be "flush free" or "no flush." Caution should be exercised regarding the use of these agents because the products covalently bind six molecules of nicotinic acid to inositol,[41] which fails to release free niacin, rendering it nonabsorbable and devoid of lipid-lowering effects.[44]

Hepatotoxicity, presenting as elevated serum transaminase values, may also occur with all formulations of niacin. However, most cases of serious hepatic toxicity have been reported with the slow-release niacin compared with the immediate-release or extended-release formulations.[45] The hepatotoxicity seen with the slow-release niacin often occurs when dosages of more than 1500 mg/day are used.[45] Hepatotoxicity with the immediate-release or extended-release niacin is considered rare. Additional adverse effects of niacin include GI discomfort, including nausea and vomiting, that appears to be dose-dependent.

An additional adverse effect of niacin is the risk of hyperglycemia in diabetic patients. The hyperglycemia associated with niacin therapy is due to insulin resistance.[41] Most of the concern regarding hyperglycemia comes from early studies that used much higher doses than used in clinical practice today. The extended-release preparation increases fasting blood glucose by about 5% when administered at a dosage of 2000 mg/day.[46] Hemoglobin A1c has been shown to increase by about 0.3% when taking extended-release niacin compared with patients not on niacin therapy.[47] Because of the minimal effects on blood glucose in diabetes, niacin can be safely administered to this patient population with appropriate monitoring. Given the overall increased risk of cardiovascular events in the diabetic population and the favorable effects of niacin on the lipid profile, the risk of therapy is probably worth the potential for a small modification in glycemic control.

DRUG INTERACTIONS

There are few drug interactions with niacin. However, when combining Niaspan with cholestyramine or colestipol, a significant amount of niacin (98%) is bound by the BAS, thus preventing absorption. Therefore, at least 4–6 hours should elapse between the administration of these two agents when given as combination therapy.

DOSING AND ADMINISTRATION

Extended-release niacin should be initiated at a dose of 500 mg administered at bedtime and may be preceded by a 325-mg dose of aspirin, 200 mg of ibuprofen, or other nonsteroidal 30 minutes before to minimize flushing.[48] In addition, niacin may be administered with a low-fat snack to slow the absorption and help reduce the severity of this adverse effect. The niacin dose can then be further increased to 1000 mg daily after 4 weeks. After the first 8 weeks of therapy, if response is inadequate, the dose may be continued to be titrated to 1500 mg and then to 2000 mg if needed. Doses greater than 2000 mg have been used safely, but may increase the risk of hepatotoxicity. The titration rate should not exceed 500 mg every 4 weeks. If a patient discontinues therapy for 3 or more days, retitration of the drug is recommended. Extended-release niacin should be used with caution in patients with a history of liver disease, those who consume excessive amounts of alcohol, those with unexplained hepatic transaminase elevations, those with gouty arthritis, or those with renal disease. In addition, the use of extended-release niacin is contraindicated in patients with active peptic ulcer disease or arterial bleeding. Substitution of OTC sustained-release niacin for the immediate-release formulation may result in severe hepatotoxicity. All patients taking niacin should have liver function tests performed at baseline and every 6–12 weeks for the first year of therapy and then every 6 months.

Immediate-release and slow-release niacin should be dosed 2 times/day beginning at 125–250 mg twice daily and titrated upward to a maximum dose of 2000 mg/day.

OMEGA-3 FATTY ACIDS

Omega-3 fatty acids are polyunsaturated fatty acids that are derived from plant and marine sources. The use of omega-3 fatty acids can reduce triglyceride concentrations by as much as 50%. However, this lowering of triglycerides may also result in increased LDL-C, especially in patients with very high triglycerides (more than 500 mg/dL). There is currently only one prescription omega-3 fatty acid product on the market, Lovaza. It is the purest form of omega-3 fatty acids available and contains 85% of EPA and DHA in a 1000-mg capsule. Different OTC omega-3 fatty acid products have a variety of different EPA and DHA doses in different ratios, all of which contain a lower amount of EPA and DHA compared with Lovaza. However, if OTC products are recommended, those with the highest content of EPA/DHA should be recommended.

MECHANISM OF ACTION

The exact mechanism by which omega-3 acids lowers triglycerides has not been fully elucidated. Proposed mechanisms include inhibition of acyl CoA:1,2-diacylglycerol acyltransferase and increased peroxisomal B oxidation. Triglyceride lowering may be due to a reduction in hepatic triglyceride synthesis because DHA and EPA inhibit the esterification of other fatty acids.[49] The efficacy of omega-3 fatty acids is summarized in Table 11.

PHARMACOKINETICS/PHARMACODYNAMICS

Pharmacokinetic or pharmacodynamic data for omega-3 acids are limited. However, the omega-3 acids administered as ethyl esters (i.e., Lovaza) are well absorbed after oral administration. The OTC products are considered nutritional supplements and are not subject to FDA regulations regarding bioavailability.

ADVERSE EFFECTS

Omega-3 fatty acids are generally well tolerated, with the most common adverse effect being a "fishy" burp smell called eructation. Other adverse effects include a fishy aftertaste, belching, and GI upset, including nausea. Over-the-counter products may be refrigerated to minimize this adverse effect. However, it is not recommended that Lovaza be refrigerated.

TABLE 11. EFFECT OF PRESCRIPTION OMEGA-3 FATTY ACIDS (LOVAZA) ON LIPID PARAMETERS

	TC	LDL-C	TG	HDL-C
Lovaza 4 g day/day[a]	−9.7%	+44.5%	−44.9%	+9.1%

[a]Approximate reduction when triglycerides are more than 500 mg/dL.
HDL-C = high-density lipoprotein cholesterol; LDL-C = low-density lipoprotein cholesterol; TC = total cholesterol; TG = triglycerides.
Reliant Pharmaceuticals. Omega-3-acid ethyl esters 90 (Lovaza) [package insert]. Liberty Corner, NJ; June 2007.

DRUG INTERACTIONS

No significant drug interactions have been reported. However, some studies suggest that the use of omega-3 fatty acids prolongs bleeding time, yet this does not appear to exceed the normal range. In patients taking warfarin and omega-3 fatty acids, more frequent monitoring of the INR may be necessary.

DOSING AND ADMINISTRATION

The recommended dose of Lovaza is four capsules (each capsule contains 1 g) administered either as a single dose or divided into two doses of 2 g each for severe hypertriglyceridemia. Some clinicians may use the product at 1-g daily doses as a prophylaxis to prevent a myocardial infarction. The drug is indicated as adjunctive therapy to diet of patients with triglyceride concentrations greater than 500 mg/dL.

ADDITIONAL INFORMATION

Omega-3 fatty acids are also available OTC. The AHA recommends that patients without documented coronary artery disease consume some type of fish at least twice per week. In patients with CHD, 1 g of EPA and DHA should be consumed daily, and in patients with hypertriglyceridemia, 2–4 g of EPA and DHA daily are recommended, provided the patient is under the care of a physician. When recommending fish oil supplementation, patients must understand that OTC products differ in their EPA/DHA content. It is important to focus on the amount of omega-3 fatty acids contained in each capsule rather than on the amount of fish oil concentrate. Some products may require the consumption of up to 11 capsules to obtain the same amount of omega-3 fatty acid content that is obtained from 4 capsules of Lovaza.[50]

REFERENCES

1. Lipitor [package insert]. New York: Pfizer; 2006.
2. Fluvastatin sodium (Lescol XL) [package insert]. East Hanover, NJ: Novartis Pharmaceutical; 2006.
3. Lovastatin (Mevacor) [package insert]. Whitehouse Station, NJ: Merck & Co.; 2005.
4. Pravastatin (Pravachol) [package insert]. Princeton, NJ: Bristol-Myers Squibb; 2002.
5. Rosuvastatin (Crestor) [package insert]. Wilmington, DE: AstraZeneca Pharmaceuticals; 2005.
6. Simvastatin (Zocor) [package insert]. Whitehouse Station, NJ: Merck & Co.; 2005.
7. Chong PH. An overview of lipid management with HMG-CoA reductase inhibitors. Consult Pharm 1998;4:399–420.
8. Chong PH, Yim BT. Rosuvastatin for the treatment of patients with hypercholesterolemia. Ann Pharmacother 2002;36:93–101.
9. Grundy SM, Vega GL. Influence of mevinolin on metabolism of low density lipoproteins in primary moderate hypercholesterolemia. J Lipid Res 1985;26:1464–75.
10. Arad Y, Ramakrishnan R, Ginsberg HN. Lovastatin therapy reduces low density lipoprotein apoB levels in subjects with combined hyperlipidemia by reducing the production of apoB-containing lipoproteins: implications for the pathophysiology of apoB production. J Lipid Res 1990;31:567–82.
11. Aguilar-Salinas CA, Barrett H, Schonfeld G. Metabolic modes of action of the statins in the hyperlipoproteinemias. Atherosclerosis 1998;141:203–7.
12. Colosimo RJ, Nunn-Thompson C. HMG-CoA reductase inhibitors. P&T 1993:21–31, 65.
13. McEvoy GK, Livak K, Welsh OH, Snow EK, Dewey DR. eds. HMG-CoA reductase inhibitors: general statement. In: AHFS Drug Information 2001. Bethesda, MD: American Society of Health-System Pharmacists, 2001:1705–17.
14. Davignon J. The cardioprotective effects of statins. Curr Atheroscler Rep 2004;6:27–35.
15. Lund TM, Torsvik H, Falch D, et al. Effect of morning versus evening intake of simvastatin on the serum cholesterol level in patients with coronary artery disease. Am J Cardiol 2002;90:784–5.
16. Wallace A, Chinn D, Rubin G. Taking simvastatin in the morning compared with in the evening: randomized controlled trial. BMJ 2003;327:788.
17. McKenney JM, Davidson MH, Jacobson TA, Guyton JR. Final conclusions and recommendations of the National Lipid Association statin safety assessment task force. Am J Cardiol 2006;97:S89–S94.
18. Cohen DE, Anania FA, Chalasani N. An assessment of statin safety by hepatologists. Am J Cardiol 2006;97:77C–81C.
19. Jacobson TA. Statin safety: lessons from new drug applications for marketed statins. Am J Cardiol 2006;97:S44–S51.

20. Bays H. Statin safety: an overview and assessment of the data. Am J Cardiol 2005;97:S6–S26.
21. Kasiske BL, Wanner C, O'Neill WC. An assessment of statin safety by nephrologists. Am J Cardiol 2006;97:S82–5.
22. Davidson MH. Combination therapy for dyslipidemia: safety and regulatory considerations. Am J Cardiol 2002;90(suppl):50K–60K.
23. Kyrklund C, Backman JT, Kivisto KT, et al. Plasma concentrations of active lovastatin acid are markedly increased by gemfibrozil but not by bezafibrate. Clin Pharmacol Ther 2001;69:340–5.
24. Ezetimibe (Zetia) [package insert]. North Wales, PA: Merck/Schering-Plough Pharmaceuticals; 2006.
25. Cholestyramine (Questran) [package insert]. Princeton, NJ: Bristol-Myers Squibb; 1997.
26. Colestipol (Colestid) [package insert]. Kalamazoo, MI: Pharmacia & Upjohn; 1999.
27. Colesevelam hydrochloride (Welchol) [package insert]. Parsippany, NJ: Daiichi Sankyo; 2006.
28. Fenofibrate (TriCor) [package insert]. North Chicago, IL: Abbott Laboratories; 1998.
29. Gemfibrozil (Lopid) [package insert]. New York: Pfizer; 2002.
30. Hottelart C, El Elper N, Rose F, et al. Fenofibrate increases creatininemia by increasing metabolic production of creatinine. Nephron 2002;92:536–41.
31. Levin A, Duncan L, Djurdjev O, et al. A randomized placebo-controlled double blind trial of lipid lowering strategies in patients with renal insufficiency: diet modification with or without fenofibrate. Clin Nephrol 200;53:140–6.
32. Deighan CJ, Caslake MJ, McConnell M, et al. Comparative effects of cerivastatin and fenofibrate on the atherogenic lipoprotein phenotype in proteinuric renal disease. J Am Soc Nephrol 2001;12:341–8.
33. Keech A, Simes RJ, Barter P, et al. Effects of long-term fenofibrate therapy on cardiovascular events in 9795 people with type 2 diabetes mellitus (the FIELD study): randomized controlled trial. Lancet 2005;366:1849–61.
34. Ansquer JC, Foucher C, Rattier S, et al. Fenofibrate reduces progression to microalbuminuria over 3 years in a placebo-controlled study in type 2 diabetes: results from the Diabetes Atherosclerosis Intervention Study (DIAS). Am J Kidney Dis 2005;45:485–93.
35. Dalton RN, Crimet D, Ansquer JC. The effect of fenofibrate on glomerular filtration rate (GFR) and other renal function tests: a double-blind placebo-controlled cross-over study in healthy subjects. Philadelphia: American Society of Nephrology, 2005.
36. Hottelar C, el Esper N, Achard JM, et al. Fenofibrate increases blood creatinine, but does not change the glomerular filtration rate in patients with mild renal insufficiency. Nephrologia 1999;20:41–4.
37. Tsimihodimas V, Kakafika A, Elisaf M. Fibrate treatment can increase serum creatinine levels [Letter]. Nephrol Dial Transplant 2001;16:1031.

38. Kasiske B, Cosio FG, Betao J, et al. Clinical practice guidelines for managing dyslipidemias in kidney transplant patients. A report from the Managing Dyslipidemias in Chronic Kidney Disease Work Group of the National Kidney Foundation Kidney Disease Outcomes Quality Initiative. Am J Transplant 2004;4(suppl):13–53.
39. Davidson MH, Armani A, McKenney JM, et al. Safety considerations with fibrate therapy. Am J Cardiol 2007;99(suppl):3C–18C.
40. Niacin (Niaspan) [package insert]. Cranbury, NJ: Kos Pharmaceuticals; 2005.
41. Guyton JR, Bays HE. Safety considerations with niacin therapy. Am J Cardiol 2007;99(suppl):22C–31C.
42. Morrow JD, Awad JA, Oates JA, et al. Identification of skin as a major site of prostaglandin D2 release following oral administration of niacin in humans. J Invest Dermatol 1992;98:812–5.
43. Stern RH, Spence JD, Freeman DJ, Parbtani A. Tolerance to nicotinic acid flushing. Clin Pharmacol Ther 1991;50:66–70.
44. Meyers CD, Carr MC, Park S, Brunzell JD. Varying cost and free nicotinic acid content in over-the-counter niacin preparations for dyslipidemia. Ann Intern Med 2003;139:996–1002.
45. McKenney JM, Proctor JD, Harris S, Chinchili VM. A comparison of efficacy and toxic effects of sustained vs immediate release niacin in hypercholesterolemic patients. JAMA 1994;271:672–7.
46. Guyton JR, Goldberg AC, Kreisberg RA, et al. Effectiveness of once-nightly dosing of extended release niacin alone and in combination hypercholesterolemia. Am J Cardiol 1998;82:737–43.
47. Grundy SM, Vega GL, McGovern ME, et al.; for the Diabetes Multicenter Research Group. Efficacy, safety, and tolerability of once daily niacin for the treatment of dyslipidemia associated with type 2 diabetes: results of the assessment of diabetes control and evaluation of the efficacy of Niaspan trial. Arch Intern Med 2002;162:1568–76.
48. Oberwittler H, Baccara-Dinet M. Clinical evidence for use of acetyl salicylic acid in control of flushing related to nicotinic acid treatment. Int J Clin Pract 2006;60:707–15.
49. Reliant Pharmaceuticals. Omega-3-acid ethyl esters 90 (Lovaza) [package insert]. Liberty Corner, NJ; June 2007.
50. Bays HE. Safety considerations with omega-3 fatty acid therapy. Am J Cardiol 2007;99(suppl):35C–43C.

6

LANDMARK CLINICAL TRIALS AND OTHER RELEVANT PUBLICATIONS

Joseph J. Saseen, Pharm.D., FCCP, BCPS (AQ Cardiology), CLS, and Sarah A. Spinler, Pharm.D., FCCP, BCPS (AQ Cardiology)

Evidence-based medicine has been defined as the conscientious, explicit, and judicious use of current best evidence in making decisions about the care of individual patients.[1] Clinicians can practice evidence-based medicine when managing dyslipidemia by applying the results of relevant clinical trials and other publications to patient care. The focus of this chapter is to reference and briefly comment on the application of select landmark clinical trials and other publications that are relevant to treating patients with dyslipidemia. These publications are essential references in the management of dyslipidemia.

CLINICAL GUIDELINES AND EXPERT CONSENSUS DOCUMENTS

In 2001, the NCEP ATP III published a clinical guideline as an executive summary report.[2] This guideline is discussed in Chapter 2. Clinicians should be aware that the NHLBI sanctioned the ATP III, which is considered the preeminent clinical guideline in the United States for managing dyslipidemia. The adapted Framingham Risk Calculator (see Chapter 2), published as an appendix in this guideline, is a primary risk assessment tool. It should be used to guide the management not only of dyslipidemia, but also of hypertension and use of low dose aspirin for prevention of MI.[3] The NCEP published an update of the ATP III in 2004 in response to newer clinical trials to recommend optional LDL-C goals of less than 70 mg/dL for very high-risk patients and less than 100 mg/dL for moderately high-risk patients.[4]

The AHA/American College of Cardiology has also published clinical guidelines that recommend aggressive lipid-lowering therapy in very high-risk patients. In 2006, the guidelines for secondary prevention for patients with CHD and other forms of AVD identified an LDL-C of less than 70 mg/dL as a "reasonable" goal for patients with any form of AVD.[5] These guidelines are endorsed by the NHLBI and strongly support the ATP III LDL-C goal of less than 70 mg/dL. The National Kidney Foundation has published the K/DOQI clinical guidelines for patients with kidney disease.[6] Under these guidelines, patients with moderate CKD are considered CHD risk equivalents, with an LDL-C goal of less than 100 mg/dL. This is in contrast to the ATP III guidelines, in which CKD is not yet classified as a CHD risk equivalent, owing to an overall paucity of clinical trials demonstrating reduced risk of cardiovascular events with lipid-lowering therapy in this population.

The National Lipid Association has commissioned several task force groups to conduct systematic, evidence-based evaluations regarding the safety of lipid-lowering drugs. These task forces have published findings for statins,[7] fibric acid derivatives,[8] niacin,[9] omega-3 fatty acids,[10] and gastrointestinally active drugs.[11] Of note, the statin safety assessment task force paper provides contemporary definitions of statin-associated muscle toxicity (myopathy and rhabdomyolysis) and provides clinical recommendations for evaluating patients with muscle symptoms while on statin agents.[7]

CLINICAL TRIALS EVALUATION: CARDIOVASCULAR END POINTS

PRIMARY PREVENTION

Three placebo-controlled clinical trials have demonstrated reductions in cardiovascular events with statin-based therapy (see Table 1).[12-14] The WOSCOPS (West of Scotland Coronary Prevention Study Group) is considered a landmark trial because it was the first statin-based clinical trial to show long-term event reductions.[12] Most patients in the AFCAPS/TexCAPS (Air Force/Texas Coronary Atherosclerosis Prevention Study), most of whom had low HDL-C, had only one or two major CHD risk factors. Therefore, the AFCAPS/TexCAPS was considered a landmark trial because it highlighted the importance of treating primary prevention patients with baseline LDL-C values that were much closer to recommended goal values than the WOSCOPS.[13]

The most compelling evidence supporting an LDL-C goal of less than 100 mg/dL for moderately high-risk patients (primary prevention patients with Framingham Risk Scores of 10–20%) is from the ASCOT-LLA (Anglo-Scandinavian Cardiac Outcomes Trial—Lipid Lowering Arm trial.[14] According to 2001 ATP III recommendations, these patients had LDL-C goals of less than 130 mg/dL, and most would not have been treated with drug therapy. However, patients treated with statin therapy attained a mean LDL-C of 90 mg/dL, which resulted in reduced cardiovascular events and provided evidence to justify the 2004 ATP III updated optional LDL-C goal of less than 100 mg/dL in moderately high-risk patients.

SECONDARY PREVENTION

Cumulatively, the 4S (Scandinavian Simvastatin Survival Study), LIPID (Long-Term Intervention with Pravastatin in Ischaemic Disease), and CARE (Cholesterol and Recurrent Events) trials are landmark trials providing clear evidence for LDL-C lowering in patients with CHD having moderate to high LDL-C values.[15-17] Results of these

TABLE 1. PRIMARY PREVENTION CLINICAL TRIALS

Clinical Trial (year)	Design	Population	Mean Baseline LDL-C (mg/dL)[a]	Mean Treatment LDL-C (mg/dL)[a]	LDL-C ↓ (%)[a]	Primary End Point and Benefit
WOSCOPS[12] (1995)	Randomized, double-blind, controlled Pravastatin 40 mg daily vs. placebo Mean 4.9 years	6595 men	192	142	26	Fatal CHD or nonfatal MI; 31% lower with pravastatin (5.5% vs. 7.9%; p<0.001)
AFCAPS/TexCAPS[13] (1998)	Randomized, double-blind, controlled Lovastatin 20–40 mg daily vs. placebo Mean 5.9 years	6605 patients	150	115	25	First acute major coronary event (fatal or nonfatal MI, unstable angina, or sudden death); 37% lower with lovastatin (6.8% vs. 10.9%; p<0.001)
ASCOT-LLA[14] (2003)	Randomized, double-blind, controlled Atorvastatin 10 mg daily vs. placebo Mean 3.3 years (stopped early because of benefit)	10,305 patients	133	90	33	Fatal CHD or nonfatal MI; 36% lower with atorvastatin (1.9% vs. 3.0%; p=0.0005)

[a] In the statin-treated group unless otherwise stated.
CHD = coronary heart disease; MI = myocardial infarction; TG = triglycerides.

trials are summarized in Table 2. Moreover, the 4S trial established that mortality is reduced in this population, of note, the 4S population had the highest baseline LDL-C values and likely represented a relatively high-risk secondary prevention population. These three trials provided the rationale for lowering LDL-C to at least less than 100 mg/dL in patients with CHD.

Three subsequent clinical trials, the TNT (Treating to New Targets), IDEAL (Incremental Decrease in End Points through Aggressive Lipid Lowering), and SPARCL (Stroke Prevention by Aggressive Reduction in Cholesterol Levels), have evaluated more aggressive LDL-C lowering with statin therapy in patients with CHD or stroke.[18-20] The TNT and IDEAL trials evaluated patients with CHD, but only the TNT trial demonstrated cardiovascular benefit based on patients' primary outcomes with more aggressive lipid lowering. The TNT provides evidence that supports an LDL-C goal of less than 70 mg/dL in stable CHD patients, as recommended by the ATP III 2004 update as an optional goal and the 2006 AHA scientific statement as a reasonable goal.[4,5] Of note, most patients in TNT had on-treatment LDL-C values in the 70s versus less than 70 mg/dL. The IDEAL trial may not have demonstrated a difference in their primary outcome because the difference in mean LDL-C values between the two treatment groups, in contrast to the TNT, was not as large. However, the IDEAL study demonstrated significant benefits with more aggressive lipid lowering on secondary outcomes, especially when stroke was included in the end point. The SPARCL trial evaluated placebo versus maximum-dose atorvastatin in patients with ischemic stroke or transient ischemic attack. At a minimum, this trial provided evidence supporting aggressive lipid lowering in this population. However, because the comparator was not active treatment, these data do not answer the question of whether an LDL-C goal less than 70 mg/dL is better than a goal less than 100 mg/dL in patients with a history of ischemic stroke or transient ischemic attack. Moreover, there was a higher incidence of hemorrhagic stroke in patients treated with atorvastatin versus placebo. Although the exact implications of this finding are debatable, this increase in risk was most apparent in patients with a history of hemorrhagic stroke on study entry.

MIXED POPULATION TRIALS

Both the ALLHAT-LLT (Antihypertensive and Lipid-Lowering Treatment to Prevent Heart Attack Trial) and HPS (Heart Protection Study) enrolled primary and secondary prevention patients (Table 3).[21,22] The ALLHAT-LLT was the only large-scale, long-term, outcome-based trial with statin therapy not to show reductions in CVD. This trial was not blinded and has been criticized because it enrolled only half of the target population and because the usual care comparison group had a larger than expected reduction in LDL-C because of its use of statin therapy outside protocol. Alternatively, the HPS is arguably the most robust and important trial for statin therapy because it was designed to evaluate several subpopulations independently. The HPS findings further reinforced the benefits of statin therapy in patients with CHD, but most importantly, they provided the first definitive evidence demonstrating cardiovascular event lowering with statin therapy in a broader high-risk range of patients

TABLE 2. SECONDARY PREVENTION CLINICAL TRIALS

Clinical Trial (year)	Design	Population	Mean Baseline LDL-C (mg/dL)[a]	Mean Treatment LDL-C (mg/dL)	LDL-C ↓ (%)[a]	Primary End Point and Benefit
4S[15] (1994)	Randomized, double-blind Simvastatin 20–40 mg daily vs. placebo Mean 5.4 years	4444 patients with CHD	187	121	35	Total mortality; 30% lower with simvastatin (8.2% vs. 11.5%; p=0.0003)
CARE[16] (1996)	Randomized, double-blind Pravastatin 40 mg daily vs. placebo Mean 5.0 years	4159 patients with prior MI	139	97	32	Fatal CHD or nonfatal MI; 24% lower with pravastatin (10.2% vs. 13.2%; p=0.003)
LIPID[17] (1998)	Randomized, double-blind Pravastatin 40 mg daily vs. placebo Mean 6.1 years	9014 patients with CHD	150	113	25	Fatal CHD; 24% lower with pravastatin (6.4% vs. 8.3%; p<0.001)
TNT[18] (2005)	Randomized Atorvastatin 80 mg daily vs. atorvastatin 10 mg daily Mean 4.9 years	10,001 patients with stable CHD and LDL-C < 130 mg/dL	98 (baseline was on atorvastatin 10 mg daily)	77 (80 mg); 101 (10 mg)	N/A	Composite end point of cardiovascular death, nonfatal MI, resuscitated cardiac arrest, and stroke; 22% lower with atorvastatin 80 mg (8.7% vs. 10.9%; p<0.001) Secondary end point: no difference in mortality (5.6% vs. 5.7%; p=0.92)

Clinical Trial (year)	Design	Population	Mean Baseline LDL-C (mg/dL)[a]	Mean Treatment LDL-C (mg/dL)	LDL-C ↓ (%)[a]	Primary End Point and Benefit
IDEAL[19] (2005)	Randomized Atorvastatin 80 mg daily vs. simvastatin 20 mg daily Mean 4.8 years	8888 patients with prior MI	121 (baseline was on statin therapy)	104 (simvastatin); 81 (atorvastatin)	N/A	Major coronary events (coronary death, nonfatal MI, resuscitated cardiac arrest) not different (9.3% vs. 10.4%; p=0.07) Secondary end point: nonfatal MI 17% lower with atorvastatin 80 mg (6.0% vs. 7.2%; p=0.02); major coronary events or stroke 13% lower with atorvastatin (12.0% vs. 13.7%; p=0.02); no difference in total mortality (8.4% vs. 8.2%; p=0.81)
SPARCL[20] (2006)	Randomized, double-blind Atorvastatin 80 mg once daily vs. placebo; mean 4.9 years	4731 patients with ischemic stroke or transient ischemic attack within 6 months and LDL-C 100–190 mg/dL and no CHD	133	73	45	Fatal or nonfatal stroke; 16% lower with atorvastatin (11.2% vs. 13.1%; p=0.03) Secondary end points: cardiovascular death, nonfatal MI, or resuscitation from cardiac arrest 35% lower with atorvastatin (3.4% vs. 5.1%; p=0.003); no difference in total mortality (9.1% vs. 8.9%; p=0.98)

[a] In the statin-treated group unless otherwise stated.

CHD = coronary heart disease; MI = myocardial infarction; TG = triglycerides.

TABLE 3. MIXED-POPULATION (PRIMARY AND SECONDARY PREVENTION) CLINICAL TRIALS

Clinical Trial (year)	Design	Population	Mean Baseline LDL-C (mg/dL)[a]	Mean Treatment LDL-C (mg/dL)[a]	LDL-C ↓ (%)[a]	Primary End Point and Benefit
ALLHAT-LLT[21] (2002)	Randomized, nonblinded Pravastatin 40 mg vs. usual care Mean 4.8 years	10,355 patients age 55 years and older with hypertension and at least one additional CHD risk factor	146	106 (pravastatin); 130 (usual care)	28 (pravastatin); 11 (usual care)	No difference in all-cause mortality (14.9% vs. 15.3%; p=0.88); 17% of patients in usual care were receiving statins by year 4 Secondary end point: no difference in fatal CHD or nonfatal MI (9.3% vs. 10.4%; p=0.16)
HPS[22] (2002)	Randomized, double-blind Simvastatin 40 mg daily vs. placebo Mean 5 years	20,536 high-risk patients with a history of CHD or noncoronary atherosclerotic vascular disease or diabetes	132	93	30	Total mortality; 13% lower with simvastatin (12.9% vs. 14.7% p=0.0003) Similar and significant reductions with simvastatin seen in fatal or nonfatal vascular events for all subcategories (cerebrovascular disease without CHD, peripheral artery disease without CHD, diabetes, women, age ≥ 70 years, baseline LDL-C below 116 mg/dL)

[a] In the statin-treated group unless otherwise stated.
CHD = coronary heart disease; MI = myocardial infarction; TG = triglycerides.

(i.e., those with stroke, those with peripheral arterial disease, the elderly, those with diabetes mellitus, and high-risk patients with lower baseline LDL-C values).

SPECIAL POPULATIONS

Several clinical trials have provided evidence evaluating the long-term effects of lipid-lowering therapy in special populations (see Table 4). In addition to the HPS, the PROSPER (PROspective Study of Pravastatin in the Elderly at Risk) demonstrated reduced cardiovascular events in an exclusively elderly population, further justifying LDL-C lowering in the elderly.[23]

Diabetes Mellitus with or without End-Stage Kidney Disease

For patients with diabetes, the CARDS (Collaborative Atorvastatin Diabetes Study), FIELD (Fenofibrate Intervention and Event Lowering in Diabetes), and 4D (German Diabetes and Dialysis Study) trials have evaluated lipid-lowering therapy.[24-26] The CARDS was the first prospective long-term trial evaluating cardiovascular event reduction in exclusively primary prevention patients with type 2 diabetes; it justified the use of statin therapy in this population.[24] The average reduction in LDL-C for patients treated with atorvastatin in this study was 40%, and the average LDL-C during treatment in most patients was in the 70s. Therefore, these data might be used to justify more aggressive LDL-C lowering than the standard LDL-C goal of less than 100 mg/dL in primary prevention patients with type 2 diabetes.

Unfortunately, the FIELD and 4D trials both failed to meet their primary end point.[25,26] The FIELD trial evaluated fenofibrate and may have failed to demonstrate a reduction in CHD events because of a significantly higher rate of statin use in the placebo-treated patients compared with fenofibrate. Of note, the secondary end point of total cardiovascular events was reduced. Nonetheless, the long-term benefits of fibrate therapy in type 2 diabetes are still unclear, which is in sharp contrast to the strong evidence supporting statin therapy in type 2 diabetes. In the 4D, patients with type 2 diabetes who were undergoing hemodialysis for kidney failure did not experience reductions in the primary end point of cardiovascular events. Similar to FIELD, there were reductions in other secondary end points, but these results are disappointing considering that a moderate dose of atorvastatin was used, a large degree of LDL-C lowering was achieved, and it was anticipated that positive outcomes that are more positive would be demonstrated. The AURORA study (a randomized, double-blind, placebo-controlled study) will provide more definitive data regarding the effect of statin therapy on the incidence of cardiovascular events in patients on hemodialysis.

Asian Patients

The MEGA (Management of Elevated Cholesterol in the Primary Prevention Group of Adult Japanese) trial was the first large-scale trial to demonstrate cardiovascular event reduction with statin therapy in an exclusively Asian population (Japanese).[27] This trial demonstrated benefits even though a low dose of statin was used, and only a minimal reduction in LDL-C was attained. These data extend the benefit to a broader range of patients, particularly to a different racial group of patients who may exhibit different pharmacokinetic responses when compared with white patients.

TABLE 4. CLINICAL TRIALS IN SPECIAL POPULATIONS

Clinical Trial (year)	Design	Population	Mean Baseline LDL-C (mg/dL)[a]	Mean Treatment LDL-C (mg/dL)	LDL-C ↓ (%)[a]	Primary End Point and Benefit
PROSPER[23] (2002)	Randomized, double-blind Pravastatin 40 mg daily vs. placebo Mean 3.2 years	5804 patients age 70–82 years; primary and secondary prevention	147	107	27	Fatal CHD, nonfatal MI, or nonfatal stroke 15% lower with pravastatin (14.1% vs. 16.2%; p=0.014)
CARDS[24] (2004)	Randomized, double-blind Atorvastatin 10 mg daily vs. placebo Mean 3.9 years (stopped early because of benefit)	2838 patients with type 2 diabetes and no history of CHD	117[b]	71[b]	40	First occurrence: an acute CHD event, coronary revascularization, or stroke 37% lower with atorvastatin (5.8% vs. 9.0%; p=0.001)
FIELD[25] (2005)	Randomized, double-blind Fenofibrate 200 mg vs. placebo (other lipid-lowering drug utilized) Mean 5 years	9795 patients with type 2 diabetes mellitus	LDL-C 118, HDL-C 42.5, TG 172	Variable[c] fenofibrate: LDL-C 94, HDL-C 44, TG 130 Placebo: LDL-C 101, HDL-C 43, TG 166	N/A	No difference in CHD events (5.3% vs. 5.9%; p=0.16) Secondary end points: total cardiovascular events lower with fenofibrate (13% vs. 14%; p=0.035); no difference in cardiovascular mortality (5.1% vs. 5.5%; p=0.41) More patients randomized to placebo started statin during study than fenofibrate (17% vs. 8%; p<0.0001)
4D[26] (2005)	Randomized, double-blind Atorvastatin 20 mg[d] daily vs. placebo Mean 4 years	1255 patients with type 2 diabetes mellitus undergoing hemodialysis	121[b] (atorvastatin); 125 (placebo)	72[b]	42	No difference in CHD death, nonfatal MI, or stroke (37% vs. 38%; p=0.37) Secondary end points: No difference in total mortality (48% vs. 50%; p=0.33); total cardiac events lower with atorvastatin (33% vs. 39%; p=0.03)

Clinical Trial (year)	Design	Population	Mean Baseline LDL-C (mg/dL)[a]	Mean Treatment LDL-C (mg/dL)	LDL-C ↓ (%)[a]	Primary End Point and Benefit
MEGA[27] (2006)	Randomized, open-label, blinded end point evaluation Pravastatin 10–20 mg daily with diet vs. diet alone Mean 5.3 years	7832 Japanese patients without CHD	157	128 (pravastatin plus diet); 151 (diet alone)	18 (pravastatin plus diet); 3.2 (diet alone)	CHD events lower in the diet plus pravastatin group than in the diet-alone group (3.3% vs. 5%; p=0.01)
PROVE-IT[30] (2004)	Randomized, double-blind Atorvastatin 80 mg daily vs. pravastatin 40 mg daily Mean 2 years	4162 patients within 10 days of an acute coronary syndrome	106[b,e]	62[b] (atorvastatin); 95[b] (pravastatin)	41[e] (atorvastatin); 10[e] (pravastatin)	All-cause mortality, acute coronary syndrome, revascularization, or stroke lower with atorvastatin (22.4% vs. 26.3%; p=0.005)
A to Z[31] (2004)	Randomized, double-blind Intense simvastatin (40 mg daily for 1 month; then 80 mg daily) vs. delayed conservative simvastatin (placebo for 4 months; then 20 mg daily) Mean 2 years	4497 patients within 5 days of an acute coronary syndrome	112[b,e] (intense); 111[b,e] (delayed conservative)	At 2 years: 66[b] (intense); 81[b] (delayed conservative)	41[d] (intense); 27[e] (delayed conservative)	No difference in combined end point of cardiovascular death, nonfatal MI, acute coronary syndrome, or stroke (14.4% vs. 16.7%; p=0.14) Myopathy (CK > 10 times ULN) higher with intense simvastatin (0.4% vs. 0.1%; p=0.02)

[a]In the statin-treated group unless otherwise stated.
[b]Data represented as median values.
[c]Variable changes in lipoproteins throughout the study because of high use of other lipid-lowering agents in both groups.
[d]Dose reduced by 50% in patients with LDL-C of less than 50 mg/dL.
[e]Falsely low because baseline LDL-C was measured during the initial acute coronary syndrome.
CHD = coronary heart disease; MI = myocardial infarction; TG = triglycerides.

Acute Coronary Syndromes

Lipid-lowering therapy for patients soon after an acute coronary syndrome is accepted as a standard of therapy.[28,29] The PROVE-IT TIMI 22 trial (Pravastatin or Atorvastatin Evaluation and Infection Therapy—Thrombolysis in Myocardial Infarction) is considered a landmark trial that demonstrated superior cardiovascular event lowering when intense statin therapy was implemented within 10 days of acute coronary syndrome versus moderate-intensity statin therapy.[30] These data were a major impetus for the ATP III 2004 update recommendation of an optional LDL-C goal of less than 70 mg/dL in very high-risk patients. In contrast, the A to Z (Aggrastat to Zocor) trial failed in its attempt to show a reduction in its primary end point when intense statin therapy was compared with delayed conservative therapy in patients with acute coronary syndrome.[31] Almost none of the secondary and tertiary end points demonstrated a difference; nevertheless, a trend suggesting a better cardiovascular event reduction with intense therapy was eventually seen. In comparison with the PROVE-IT, the difference in LDL-C between the two comparison groups was much smaller in the A to Z, which might explain the discrepancy between the results of these two trials. The A to Z also demonstrated a higher risk of myopathy with simvastatin 80 mg versus lower doses, suggesting that simvastatin 80 mg has a significantly higher risk of serious muscle toxicity than lower, safer doses of this agent (i.e., simvastatin 40 mg daily that was used in HPS).

CLINICAL TRIALS ASSESSING SURROGATE MARKERS OF ATHEROSCLEROSIS

Three large-scale prospective clinical trials evaluating atheroma burden as a surrogate marker have been conducted (see Table 5).[32-34] Intravascular ultrasound was used in the REVERSAL (Reversal of Atherosclerosis with Aggressive Lipid Lowering) and ASTEROID (Study to Evaluate the Effect of Rosuvastatin on Intravascular Ultrasound-Derived Coronary Atheroma Burden) trials to measure coronary atheroma burden.[31,32] B-mode ultrasound was used in METEOR (Measuring Effects on Intima-Media Thickness: An Evaluation of Rosuvastatin) to measure carotid atheroma.[33] These methods, which are used to measure surrogate markers of clinical AVD, can provide parameters that suggest atherosclerotic progression, stabilization, or regression. Lipid lowering reduces cardiovascular events in a broad range of populations. Conducting long-term placebo-controlled trials evaluating the effects of a lipid-lowering drug on cardiovascular events is complicated and, in many instances, considered unethical. Therefore, these surrogate measures are used to demonstrate the speculative long-term benefits with newer lipid-lowering agents or newer approaches to treating CVD. These three trials with statins suggest that intense lipid lowering halts and/or regresses clinical AVD. Although they will always be criticized as are surrogate markers, their use is increasing, and results of these trials provide evidence supporting aggressive lipid lowering in certain patients.

TABLE 5. CLINICAL TRIALS ASSESSING SURROGATE MARKERS OF ATHEROSCLEROSIS

Clinical Trial (year)	Design	Population	Mean Baseline LDL-C (mg/dL)[a]	Mean Treatment LDL-C (mg/dL)[a]	LDL-C ↓ (%)[a]	Effect on Atheroma Volume
REVERSAL[33] (2004)	Randomized, double-blind Pravastatin 40 mg daily vs. atorvastatin 80 mg for 18 months IVUS used to measure coronary atheroma volume	654 patients with at least one coronary artery with at least a 20% stenosis on coronary angiography	150	110 pravastatin, 79 atorvastatin	25 pravastatin, 46 atorvastatin	Progression of coronary atherosclerosis from baseline in 2.7% in the pravastatin group ($p=0.001$); progression did not occur in the atorvastatin group ($p=0.98$)
ASTEROID[32] (2006)	Open-label, blinded end point Rosuvastatin 40 mg for 24 months IVUS used to measure coronary atheroma volume	507 patients with at least one coronary artery with at least a 20% stenosis on coronary angiography	130 (HDL-C 43)	61 (HDL-C 49)	53 (HDL-C 14.7% ↑)	Significant reductions in percent atheroma volume and change in nominal atheroma volume in the 10-mm subsegment with the greatest disease severity compared with baseline ($p<0.001$ for both)
METEOR[34] (2007)	Randomized, double-blind Rosuvastatin 40 mg vs. placebo for 24 months B-mode ultrasound to measure carotid intima-media thickness	876 patients at low risk of CHD (98% Framingham Risk Score < 10%) and modest carotid intima-media thickness	155	78	49	Significant reduction in the rate of progression with carotid intima-media thickness with rosuvastatin vs. placebo ($p<0.001$) but no regression demonstrated with rosuvastatin

[a] In the statin-treated group unless otherwise stated.
IVUS = intravascular ultrasound.

REFERENCES

1. Sackett DL, Rosenberg WM, Gray JA, et al. Evidence based medicine: What it is and what it isn't. BMJ 1996;312:71–2.
2. Executive Summary of the Third Report of the National Cholesterol Education Program (NCEP) Expert Panel on Detection, Evaluation, and Treatment of High Blood Cholesterol in Adults (Adult Treatment Panel III). JAMA 2001;285:2486–97.
3. Rosendorff C, Black HR, Cannon CP, et al. Treatment of hypertension in the prevention and management of ischemic heart disease: A scientific statement from the American Heart Association Council for High Blood Pressure Research and the Councils on Clinical Cardiology and Epidemiology and Prevention. Circulation 2007;115:2761–88.
4. Grundy SM, Cleeman JI, Merz CN, et al. Implications of recent clinical trials for the National Cholesterol Education Program Adult Treatment Panel III guidelines. Circulation 2004;110:227–39.
5. Smith SC Jr, Allen J, Blair SN, et al. AHA/ACC guidelines for secondary prevention for patients with coronary and other atherosclerotic vascular disease: 2006 update endorsed by the National Heart, Lung, and Blood Institute. J Am Coll Cardiol 2006;47:2130–9.
6. K/DOQI clinical practice guidelines for management of dyslipidemias in patients with kidney disease. Am J Kidney Dis 2003;41(4 suppl 3):I–IV, S1–91.
7. McKenney JM, Davidson MH, Jacobson TA, Guyton JR. Final conclusions and recommendations of the National Lipid Association Statin Safety Assessment Task Force. Am J Cardiol 2006;97:89C–94C.
8. Davidson MH, Armani A, McKenney JM, Jacobson TA. Safety considerations with fibrate therapy. Am J Cardiol 2007;99:3C–18C.
9. Guyton JR, Bays HE. Safety considerations with niacin therapy. Am J Cardiol 2007;99:22C–31C.
10. Bays HE. Safety considerations with omega-3 fatty acid therapy. Am J Cardiol 2007;99:35C–43C.
11. Jacobson TA, Armani A, McKenney JM, Guyton JR. Safety considerations with gastrointestinally active lipid-lowering drugs. Am J Cardiol 2007;99:47C–55C.
12. Shepherd J, Cobbe SM, Ford I, et al. Prevention of coronary heart disease with pravastatin in men with hypercholesterolemia. West of Scotland Coronary Prevention Study Group. N Engl J Med 1995;333:1301–7.
13. Downs JR, Clearfield M, Weis S, et al. Primary prevention of acute coronary events with lovastatin in men and women with average cholesterol levels: results of AFCAPS/TexCAPS. Air Force/Texas Coronary Atherosclerosis Prevention Study. JAMA 1998;279:1615–22.
14. Sever PS, Dahlof B, Poulter NR, et al. Prevention of coronary and stroke events with atorvastatin in hypertensive patients who have average or lower-than-average cholesterol concentrations, in the Anglo-Scandinavian Cardiac Outcomes Trial—Lipid Lowering Arm (ASCOT-LLA): a multicentre randomised controlled trial. Lancet 2003;361:1149–58.

15. Randomised trial of cholesterol lowering in 4444 patients with coronary heart disease: the Scandinavian Simvastatin Survival Study (4S). Lancet 1994;344:1383–9.
16. Sacks FM, Pfeffer MA, Moye LA, et al. The effect of pravastatin on coronary events after myocardial infarction in patients with average cholesterol levels. Cholesterol and Recurrent Events Trial investigators. N Engl J Med 1996;335:1001–9.
17. The Long-Term Intervention with Pravastatin in Ischaemic Disease (LIPID) Study Group. N Engl J Med 1998;339:1349–57. Prevention of cardiovascular events and death with pravastatin in patients with coronary heart disease and a broad range of initial cholesterol levels.
18. LaRosa JC, Grundy SM, Waters DD, et al. Intensive lipid lowering with atorvastatin in patients with stable coronary disease. N Engl J Med 2005;352:1425–35.
19. Pedersen TR, Faergeman O, Kastelein JJ, et al. High-dose atorvastatin vs usual-dose simvastatin for secondary prevention after myocardial infarction: the IDEAL study: a randomized controlled trial. JAMA 2005;294:2437–45.
20. Amarenco P, Bogousslavsky J, Callahan A 3rd, et al. High-dose atorvastatin after stroke or transient ischemic attack. N Engl J Med 2006;355:549–59.
21. Major outcomes in moderately hypercholesterolemic, hypertensive patients randomized to pravastatin vs usual care: the Antihypertensive and Lipid-Lowering Treatment to Prevent Heart Attack Trial (ALLHAT-LLT). JAMA 2002;288:2998–3007.
22. Heart Protection Study Collaborative Group. MRC/BHF Heart Protection Study of cholesterol lowering with simvastatin in 20,536 high-risk individuals: a randomised placebo-controlled trial. Lancet 2002;360:7–22.
23. Shepherd J, Blauw GJ, Murphy MB, et al. Pravastatin in elderly individuals at risk of vascular disease (PROSPER): a randomised controlled trial. Lancet 2002;360:1623–30.
24. Colhoun HM, Betteridge DJ, Durrington PN, et al. Primary prevention of cardiovascular disease with atorvastatin in type 2 diabetes in the Collaborative Atorvastatin Diabetes Study (CARDS): multicentre randomised placebo-controlled trial. Lancet 2004;364:685–96.
25. Keech A, Simes RJ, Barter P, et al. Effects of long-term fenofibrate therapy on cardiovascular events in 9795 people with type 2 diabetes mellitus (the FIELD study): randomised controlled trial. Lancet 2005;366:1849–61.
26. Wanner C, Krane V, Marz W, et al. Atorvastatin in patients with type 2 diabetes mellitus undergoing hemodialysis. N Engl J Med 2005;353:238–48.
27. Nakamura H, Arakawa K, Itakura H, et al. Primary prevention of cardiovascular disease with pravastatin in Japan (MEGA Study): a prospective randomised controlled trial. Lancet 2006;368:1155–63.
28. Antman EM, Anbe DT, Armstrong PW, et al. ACC/AHA guidelines for the management of patients with ST-elevation myocardial infarction—executive summary: a report of the American College of Cardiology/American Heart Association Task Force on Practice Guidelines (Writing Committee to Revise the 1999 Guidelines for the Management of Patients with Acute Myocardial Infarction). Circulation 2004;110:588–636.

29. Anderson JL, Adams CD, Antman EM, et al. ACC/AHA 2007 guidelines for the management of patients with unstable angina/non ST-elevation myocardial infarction: A report of the American College of Cardiology/American Heart Association Task Force on Practice Guidelines (Writing Committee to Revise the 2002 Guidelines for the Management of Patients with Unstable Angina/Non ST-Elevation Myocardial Infarction) developed in collaboration with the American College of Emergency Physicians, the Society for Cardiovascular Angiography and Interventions, and the Society of Thoracic Surgeons: endorsed by the American Association of Cardiovascular and Pulmonary Rehabilitation and the Society for Academic Emergency Medicine. Circulation 2007;116:803–77.
30. Cannon CP, Braunwald E, McCabe CH, et al. Intensive versus moderate lipid lowering with statins after acute coronary syndromes. N Engl J Med 2004;350:1495–504.
31. De Lemos JA, Blazing MA, Wiviott SD, et al. Early intensive vs a delayed conservative simvastatin strategy in patients with acute coronary syndromes: phase Z of the A to Z trial. JAMA 2004;292:1307–16.
32. Nissen SE, Nicholls SJ, Sipahi I, et al. Effect of very high-intensity statin therapy on regression of coronary atherosclerosis: the ASTEROID trial. JAMA 2006;295:1556–65.
33. Nissen SE, Tuzcu EM, Schoenhagen P, et al. Effect of intensive compared with moderate lipid-lowering therapy on progression of coronary atherosclerosis: a randomized controlled trial. JAMA 2004;291:1071–80.
34. Crouse JR 3rd, Raichlen JS, Riley WA, et al. Effect of rosuvastatin on progression of carotid intima-media thickness in low-risk individuals with subclinical atherosclerosis: the METEOR trial. JAMA 2007;297:1344–53.

APPENDIX: CLINICAL TRIAL ACRONYM GLOSSARY

A to Z	Aggrastat to Zocor
AFCAPS/TexCAPS	Air Force/Texas Coronary Atherosclerosis Prevention Study
ALLHAT-LLT	Antihypertensive and Lipid-Lowering Treatment to Prevent Heart Attack Trial—Lipid Lowering Trial
ASCOT-LLA	Anglo-Scandinavian Cardiac Outcomes Trial—Lipid Lowering Arm
ASTEROID	Study to Evaluate the Effect of Rosuvastatin on Intravascular Ultrasound-Derived Coronary Atheroma Burden
CARDS	Collaborative Atorvastatin Diabetes Study
CARE	Cholesterol and Recurrent Events Trial
FIELD	Fenofibrate Intervention and Event Lowering in Diabetes
HPS	Heart Protection Study
IDEAL	Incremental Decrease in End Points through Aggressive Lipid Lowering
LIPID	Long-Term Intervention with Pravastatin in Ischemic Disease
MEGA	Management of Elevated Cholesterol in the Primary Prevention Group of Adult Japanese
METEOR	Measuring Effects on Intima-Media Thickness: An Evaluation of Rosuvastatin
PROSPER	PROspective Study of Pravastatin in the Elderly at Risk
PROVE-IT TIMI	Pravastatin or Atorvastatin Evaluation and Infection Therapy—Thrombolysis in Myocardial Infarction
REVERSAL	Reversal of Atherosclerosis with Aggressive Lipid Lowering
SPARCL	Stroke Prevention by Aggressive Reduction in Cholesterol Levels
TNT	Treating to New Targets
WOSCOPS	West of Scotland Coronary Prevention Study
4D	German Diabetes and Dialysis Study
4S	Scandinavian Simvastatin Survival Study

7

SPECIAL PATIENT POPULATIONS

Anthony J. Busti, Pharm.D., BCPS, CLS

PREGNANCY

BACKGROUND

Although generally not considered a major contributor to the development of CVD, several important changes occur in the lipid profile throughout pregnancy.[1,2] These short-term effects are thought to be adaptive, allowing glucose to be saved for the developing fetus. In particular, triglycerides directly increase with increasing estrogen levels. In fact, triglyceride concentrations may double or triple from baseline at the time of delivery and may remain elevated for as long as 6 weeks postpartum. The greatest concern about this elevation is the risk of pancreatitis development. However, only 4–6% of pancreatitis cases are linked to high triglycerides (most are in patients with type I or V hyperlipidemia); the remaining cases likely result from cholelithiasis.[2] A history of HIV-related lipodystrophy and protease inhibitor (PI) use independently increases the risk of significant triglyceride elevations; these cases thus warrant greater observation.[3] In addition, LDL-C concentrations decrease initially but, after about 2 months of pregnancy, can increase by 25–50% and remain elevated for as long as 8 weeks postpartum. Qualitatively, the LDL-C also changes to a smaller, denser LDL-C because of the exchange of triglycerides and cholesterol esters with the triglyceride-enriched VLDL.[4] Finally, there is an overall increase in HDL_2 compared with preterm levels but no changes in HDL_3 concentrations, which is also probably related to increasing estrogen levels. The influence and effect of repeated exposures of these quantitative and qualitative changes in the lipid profile for multiparous women at risk of atherosclerotic diseases are mostly unknown.

TABLE 1. TREATMENT CONSIDERATIONS IN PREGNANCY

Treatment	MW	PB %	Used in Pregnancy[a]	Pregnancy Category	Reviewed by AAP	Lactation Risk Catagory	M/P Ratio
HMG-CoA reductase inhibitors							
Atorvastatin	1209	98	No	X	No	L3	2.0
Fluvastatin	411	98	No	X	No	L3	
Lovastatin	404	95	No	X	No	L3	
Pravastatin	446	50	No	X	No	L3	
Rosuvastatin			No	X	No	L3	
Simvastatin	418	95	No	X	No	L3	
Cholesterol absorption inhibitors							
Ezetimibe	409	90	—	C	No	L3	
Bile acid binding resins							
Colesevelam	N/A	N/A	±	B	No	L1	0.0
Cholestyramine	N/A	N/A	±	C	No	L1	0.0
Colestipol	N/A	N/A	±	C	No		0.0
Fibric acid derivatives							
Fenofibratea	361	90	—	C	No	L3	
Gemfibrozil	250	90	±	C	No	L3	
Others							
Niacin	123		No	C	No		
Fish oil			±		No		
Plant stanols			±		No		
LDL apheresis			±		No		

Note: Empty areas reflect areas for which no data are available.

[a] Available in multiple brands with different milligram doses (TriCor Lofibra tablets, Lofibra capsules, Antara).

AAP = American Academy of Pediatrics; HMG-CoA = 3-hydroxy-3-methylglutaryl coenzyme A; LDL = low-density lipoprotein; M/P ratio = milk-plasma ratio; MW = molecular weight; N/A = not applicable; PB (%) = percent protein binding.

Lactation risk category: L1 = safest, L2 = safer, L3 = moderately safe, L4 = possible hazardous, L5 = contraindicated.

Theoretic infant dose = estimates the maximal dose the infant would receive per kilogram per day.

Relative infant dose = determined by taking the dosage of the infant (mg/kg/day) divided by the dosage in mother (mg/kg/day). In general, if < 10%, considered safe.

TREATMENT RECOMMENDATIONS

Nearly all of the prescribing information for lipid-lowering products states that they have not been studied in human pregnancy for safety and efficacy. The general recommendation is that these agents not be used or stopped before pregnancy. Similar recommendations are made regarding breastfeeding; these have not, however, been reviewed by the American Academy of Pediatrics.[5] Factors that increase the chance of a drug being secreted into breast milk include a certain degree of lipophilicity, a low degree of protein binding, and a molecular weight less than 500.[5] Overall, little is known about the use of these drugs in this special population (see Table 1).[6-20]

In general, the benefit-to-risk ratio for the treatment of hyperlipidemia in pregnancy is largely dependent on the type of dyslipidemia. It may be that women with homozygous familial hyperlipidemia (HoFH) either choose to avoid pregnancy because of the further worsening of their lipids and having to stop most lipid-lowering drugs because of safety concerns or choose to be monitored by a specialist, who may use certain techniques such as LDL-C apheresis.[1,21] It may also be prudent for such women who do decide to have children to condense the time off drugs and limit the number of childbearing years. Non-HoFH patients who are otherwise healthy will most likely be able to avoid lipid-lowering treatment during pregnancy or afterward because lipid levels return to baseline within 6 months.

STATINS

All statins have a pregnancy category X and are not recommended (including during breastfeeding).[7-12] A few studies have evaluated pregnant women who were inadvertently exposed to statins; they indicated no notable increases in congenital anomalies compared with the general population.[22-24] However, case reports have indicated central nervous system and limb anomalies with first-trimester exposure.[25]

BILE ACID BINDING RESINS

Of all the treatment options, only colesevelam has a pregnancy category rating of B.[14] Because the bile acid binding resins (BABRs) are not absorbed, fetal exposure should not be a concern. These agents also provide the greatest potential safety for lowering LDL-C. Because triglycerides can increase significantly during pregnancy, clinicians should remember that all of the BABRs increase triglycerides.[14-16,26] Finally, the BABRs interfere with fat-soluble vitamin absorption, which may be most important during pregnancy.[27-29]

CHOLESTEROL ABSORPTION INHIBITOR

Ezetimibe has a pregnancy category of C because of the observed adverse effects in animal studies; there have been no adequate studies of its use in the pregnant population.[13] Based on these studies and findings, clinicians should carefully weigh the benefits and risks of this drug before initiating therapy.

FIBRIC ACID DERIVATIVES

Even if triglyceride therapy is needed during pregnancy, there is insufficient evidence to make any recommendations. Although the fibric acid derivatives carry a pregnancy category C rating and are effective at lowering triglycerides, they also increase the risk of cholestasis that pregnant women are already at risk of developing.[30]

OMEGA-3 FATTY ACIDS AND NIACIN

No evidence exists for the use of higher doses of supplemental omega-3 fatty acids or niacin. If fish oil supplementation is to be used or if fish is to be consumed, it is prudent to use or consume sources of fish oils that contain the least amount of mercury.[31] The 2007 Report of the National Lipid Association's Safety Task Force on omega-3 fatty acids supplements states that there are not likely sufficient concentrations of mercury or excessive vitamins to pose a potential health risk; however, that risk is dependent on the purification process used by the manufacturer.[32]

PLANT STANOLS

Plant sterols or stanols, either added to foods or as supplements, are not recommended for pregnant or breastfeeding women because their safety has not been studied.[33] However, there is also no evidence that vegetarian women who ingest large amounts of naturally occurring phytosterols are at increased risk of adverse affects during pregnancy or lactation. Overall, there is little evidence for the use of lipid-lowering agents during pregnancy, and risk-to-benefit ratios should be considered.

CHILDREN

BACKGROUND

The atherosclerotic process begins in childhood, but for most children, the extent of vascular involvement is minor, and the rate of progression is generally slow. In such cases, preventive measures through the promotion of healthy lifestyles and behavior modification are the focus. However, there are high-risk pediatric cases, such as those with HoFH, in which coronary disease is evident in the first and second decades of life.[34-36] Other high-risk pediatric patients include those with type 1 and type 2 diabetes mellitus, CKD, end-stage renal disease, post–heart transplantation, Kawasaki disease, and chronic inflammatory disease as well as survivors of cancer treatment (Table 2).[37]

TREATMENT RECOMMENDATIONS

Until 2007, the old NCEP guidelines (1992) for pediatrics recommended that the screening of lipids in children and adolescents aged 2–19 years be performed selectively.[37,38] These guidelines initially recommend that if the parents are known to have total cholesterol more than 240 mg/dL, measurement of a nonfasting total cholesterol or a full-fasting lipid profile be performed. A full-fasting lipid profile should be performed if a random total cholesterol reading is more than 200 mg/dL or two measurements average more than 170 mg/dL, if there is a family history of early CVD, or if there

TABLE 2. HIGH-RISK PEDIATRIC POPULATION RISK STRATIFICATION AND LIPID TREATMENT

STEP 1: Risk stratification

Tier I: High risk	Homozygous FH, type 1 diabetes mellitus, CKD/end-stage renal disease, postorthotopic heart transplant, Kawasaki disease with current artery aneurysms
Tier II: Moderate risk	Heterozygous FH, chronic inflammatory disease, Kawasaki disease with regressed aneurysms, type 2 diabetes mellitus
Tier III: At risk	Congenital heart disease, Kawasaki disease without coronary involvement, cancer treatment survivors

STEP 2: Assess other risk factors (if ≥ 2 additional: move to next higher tier)

Fasting lipid profile, smoking history, family history of early coronary artery disease in expanded 1st degree pedigree (M ≤ 55; F ≤ 65 years), BP on three separate occasions interpreted for age/sex/height, determine BMI, FG, physical activity history

STEP 3: Cut points and treatment goals

Tier I: High risk	Cut points (BMI ≤ 85th percentile for age/sex and BP ≤ 90th percentile for age/sex) Treatment goals (LDL ≤ 100 mg/dL and FG < 100 mg/dL, hemoglobin A1c < 7%)
Tier II: Moderate risk	Cut points (BMI ≤ 90th percentile for age/sex and BP ≤ 95th percentile for age/sex) Treatment goals (LDL ≤ 130 mg/dL and FG < 100 mg/dL, hemoglobin A1c < 7%)
Tier III: At risk	Cut points (BMI ≤ 85th percentile for age/sex and BP ≤ 90th percentile for age/sex) Treatment goals (LDL ≤ 100 mg/dL and FG < 100 mg/dL, hemoglobin A1c < 7%)

STEP 4: Lifestyle changes

See Table 4

STEP 5: Drug therapy for lipids

See Table 5

BMI = body mass index; BP = blood pressure; CKD = chronic kidney disease; FG = fasting glucose; FH = familial hyperlipidemia; LDL = low-density lipoprotein;

Source: *American Heart Association Scientific Statement: Cardiovascular Risk Reduction in High-Risk Pediatric Patients* 2006.

is a history of FH. In 2007, the AHA Atherosclerosis, Hypertension, and Obesity in Youth Committee, in collaboration with other groups, recommended that, in addition to family history, overweight and obesity be reasons for screening. In addition, they recommended that a risk factor assessment similar to that currently recommended by the NCEP ATP III be applied to children and adolescents to specifically identify those who would be at high risk when more aggressive treatment considerations were warranted (see Table 3).[36,37]

SPECIAL PATIENT POPULATIONS

TABLE 3. DRUG THERAPY FOR LIPIDS IN PEDIATRIC PATIENTS

Tier I: High risk	Homozygous FH: LDL: scheduled apheresis every 1–2 wk to maximally lower LDL + statin + cholesterol absorption inhibitor; refer to lipid specialist, assess BMI, BP, and FG: Step 1 for 6 mo; then to Step 2 if goals not met
	Type I diabetes mellitus: Intensive glucose management; hemoglobin A1c < 7%; assess BMI, fasting lipids → Step 1: manage weight, lipids for 6 mo; if goals not met → Step 2; statin if > 10 yr old
	CKD/ESRD: Assess BMI, BP, lipids, FG: Step 1 for 6 mo. If goals not met → Step 2; statin if > 10 yr old
	Post–Heart Transplant: Antirejection therapy, treatment of cytomegalovirus, heart imaging per transplant physician, assess BMI, BP, lipids, FG: Initiate Step 2 therapy + statins immediately in all patients > 1 yr old
	Kawasaki disease: Antithrombotic therapy, activity restriction, routine heart imaging per cardiologist; assess BMI, BP, lipids, FG: Initiate Step 1 management for 6 mo; if goals not met → Step 2; statin if > 10 yr old
Tiers II & III: LDL	If LDL ≥ 130 mg/dL (tier II) or > 160 mg/dL (tier III) → Step 1: Nutritionist evaluation and diet education for all: total fat < 30% and saturated fat < 7% of calories, cholesterol < 200 mg/day, avoid trans fat for 6 mo
	If repeat LDL > 130 mg/dL (tier II) or > 160 mg/dL (tier III) AND child > 10 yr old → Step 2: Start statin therapy for LDL goal of 130 mg/dL
Tiers II & III: TG	If initial TG = 150–400 mg/dL → Step 1: Nutritionist training for low simple carbohydrates, low-fat diet; if the ↑ TG is due to excess weight → referral for weight loss management and energy balance training plus activity recommendations
	If TG > 700–1000 mg/dL initially or at follow-up → Step 2: Consider fibrate or niacin if > 10 yr old and weight loss management when TG elevation due to excess weight or obesity

BMI = body mass index; BP = blood pressure; CKD = chronic kidney disease; ESRD = end-stage renal disease; FG = fasting glucose; FH = familial hyperlipidemia; LDL = low-density lipoprotein; TG = triglycerides.

Source: American Heart Association Scientific Statement: Cardiovascular Risk Reduction in High-Risk Pediatric Patients 2006.

The cornerstone in the prevention of CVD in childhood should focus primarily on lifestyle modifications and risk factor reduction. Until recently, BABRs were used most commonly, but GI adverse effects compromise adherence in many cases.[37–40] A prospective, randomized, double-blind, placebo-controlled trial is currently under way evaluating the safety and efficacy of colesevelam in the treatment of heterozygous FH (HeFH) in pediatric patients aged 10–17 years. Now that several clinical trails with statins have been completed, many of the statins have been shown to be safe and effective in children as young as 10 years (pravastatin is the only one with dosing in children older than 8 years).[41] In addition, the overall lack of adverse effects and use of FH in children 10 years and older has allowed ezetimibe to be used either as monotherapy or in combination with statins. The use of niacin has not been studied in patients younger than 21 and should be prescribed only by lipid specialists. Plant stanols and sterols have been shown to reduce LDL-C by about 15% with short-term use.[42-44] Overall, the clinician must balance the risk of treatment not only against the benefits of treatment, but also against the reduced risk of CVD (see Table 4).[7-20,37-44]

ACUTE CORONARY SYNDROME
BACKGROUND

The aggressive and time-sensitive treatment regimens for acute coronary syndrome are multifactorial. Intensive and early initiation of statin therapy within 24 hours of acute coronary syndrome is now a recommendation by the 2005 AHA/Advanced Cardiovascular Life Support.[45] This recommendation was based on nine randomized clinical trials and some smaller studies that demonstrated a reduction in the incidence of reinfarctions, stroke, coronary intervention for recurrent angina, and rehospitalization when statins were initiated within a few days after an acute coronary syndrome.[46-57] Two studies in STEMI (ST-segment elevation myocardial infarction) initiated statins within 6 hours of presentation.[53,54] In addition, two studies showed that patients who presented to the hospital with acute coronary syndrome who were already taking statins should continue taking them.[58,59]

Data from the National Registry of Myocardial Infarction 4 showed that early initiation of statins in the acute setting (first 24 hours) for acute myocardial infarction was associated with a significantly lower rate of early complications and in-hospital mortality.[60] This was reinforced by a recent meta-analysis of seven randomized clinical trials involving 15,968 patients admitted to the hospital for acute coronary syndrome and follow-up ranging from 1 to 24 months.[61] Early and aggressive statin therapy reduces the risk of death by 21% and the risk of cardiovascular death by 25% during the average 15-month follow-up. In addition, treating 90 patients would be required to prevent one excess death (see Table 5).

The benefits of early and aggressive statin use within the setting of acute coronary syndrome are not likely due to lowering of the cholesterol, but rather, to their pleiotropic effects (anti-inflammatory and/or antithrombotic effects).[62,63] This was suggested when evaluating the differences between the Aggrastat to Zocor (A to Z) and the Pravastatin or Atorvastatin Evaluation and Infection Therapy—Thrombolysis in Myocardial Infarction 22 (PROVE-IT-TIMI 22) trials in which CRP levels were significantly lower at 1 month with atorvastatin 80 mg in the PROVE-IT trial than with the less aggressive treatment approach of simvastatin 40 mg in the A to Z trial.[64] In addition, the additive benefit of both CRP and LDL-C reduction to predefined targets has been supported.[65,66] These reductions were observed more in PROVE-IT than in A to Z, again supporting the early and aggressive use of statins rather than a more conservative approach (see Table 5).[64] However, there were several other differences in these two trials such as the use of protein C and glycoprotein IIb/IIIa inhibitors and the location of treatments, which also may have influenced the degree of benefits between the studies. These differences have yet to be clearly determined.

TABLE 4. LIPID-LOWERING DRUG THERAPY IN PEDIATRIC PATIENTS

Treatment	Doses Studied	Ages (yr)	Conditions Studied	Lipid-Lowering Effect	Comments
HMG-CoA reductase inhibitors					
Atorvastatin	10, 20, 40 mg	10–17	HeFH, severe HLD	↓ TC 31%, ↓ LDL 39%	CYP3A4 substrate
Fluvastatin	20, 40, 80 mg	10–16	HeFH	↓ TC 21%, ↓ LDL 27%	CYP2C8/9, 3A4 substrate
Lovastatin	10, 20, 30, 40 mg	10–17	HeFH	↓ TC 23%, ↓ LDL 26%	CYP3A4 substrate
Pravastatin	5, 10, 20 mg	8–13	HeFH	↓ LDL 23–33%	Only statin with dosing for < 10 yr of age
	4 mg	14–18			
Rosuvastatin					
Simvastatin	5, 10, 20, 40 mg	10–17	HeFH	↓ TC 26%, ↓ LDL 37%	CYP3A4 substrate
Cholesterol absorption inhibitors					
Ezetimibe	10 mg	≥ 10	HoFH	↓ LDL 21%	Metabolized by glucuronidation interaction with fibrates and cyclosporine
Bile acid binding resins					
Colesevelam					Not studied in pediatrics
Cholestyramine	8–24 g	6–14	HeFH, FCHL	↓ TC 20%, ↓ LDL 26%	Can ↑ TG; ↑ GI side effects, may affect compliance; consider vitamin replacement
Colestipol	2–12 g	8–18	HeFH	↓ TC 17%, ↓ LDL 17%	Can ↑ TG; ↑ GI side effects, may affect compliance; consider vitamin replacement
Fibric acid derivatives					
Fenofibrate					No studies in < 18 yr of age
Gemfibrozil					No studies in < 18 yr of age

(Continued)

TABLE 4. LIPID-LOWERING DRUG THERAPY IN PEDIATRIC PATIENTS (Continued)

Treatment	Doses Studied	Ages (yr)	Conditions Studied	Lipid-Lowering Effect	Comments
Others					
Niacin					No studies in patients < 21 yr of age; use only by lipid specialist
Omega-3 fatty acids					
Plant stanols	2–3 g	2–15	HoFH, HeFH	↓ TC 11%, ↓ LDL 15%	Avoid in homozygous sitosterolemia

Note: Empty areas reflect areas where no data are available.
CYP = cytochrome P450; FCHL = familial combined hypercholesterolemia; GI = gastrointestinal; HeFH = heterozygous familial hyperlipidemia; HLD = hyperlipidemia; HMG-CoA = 3-hydroxy-3-methylglutaryl coenzyme A; HoFH = homozygous familial hyperlipidemia; LDL = low-density lipoprotein; TC = total cholesterol; TG = triglyceride.

TABLE 5. STATINS IN THE SETTING OF ACUTE CORONARY SYNDROME

Variable	MIRACL	PROVE-IT	A to Z	Meta-analysis (7 RCTs)
Statin	Atorvastatin	Atorvastatin	Simvastatin	
Starting dose	80 mg	80 mg	40/80 mg	
Primary end point (4 mo/2 yr)	↓ 16%/N/A	↓ 16%/↓ 16%	N/A/↓ 11%	N/A/↓ 21%
Cardiovascular mortality	↓ 8%	↓ 30%	↓ 25%	↓ 25%
Myocardial infarction	↓ 10%	↓ 13%	↓ 4%	↓ 10%
Recurrent unstable angina	↓ 26%	↓ 29%	↓ 1%	↓ 13%
C-reactive protein	↓ 34%	↓ 38%	↓ 26%/↓ 17%	
LDL	↓ 47%	↓ 33%	↓ 50%/↓ 18%	
Increase in liver enzymes (≥3 times the ULN)	2.5%	3.3%	0.9%	
Incidence of myopathy	0%	0%	0.4%	

A to Z = Aggrastat to Zocor trial; LDL = low-density lipoprotein; MIRACL = Myocardial Ischemia Reduction with Aggressive Cholesterol Lowering trial; N/A = not applicable; PROVE-IT = Pravastatin or Atorvastatin Evaluation and Infection Therapy—Thrombolysis in Myocardial Infarction 22 trial; RCT = randomized clinical trial; ULN = upper limit of normal.

TREATMENT RECOMMENDATIONS

Statins have been almost the only lipid-lowering drugs studied in the early management of acute coronary syndrome. As previously indicated, most of the benefits with statins occur with early and aggressive therapy (higher starting doses). Several studies started patients on maximal doses without any titration. The statin with the greatest evidence in this setting that also maintains overall safety is atorvastatin 80 mg.[47,52] Several other statins with efficacy at higher initial starting doses have been studied; these include pravastatin 40 mg, fluvastatin 80 mg, and simvastatin 40 mg initially with titration to 80 mg after 1 month. Although these data suggest that these starting doses are safe relative to their benefits, atorvastatin 80 mg in the Myocardial Ischemia Reduction with Aggressive Cholesterol Lowering (MIRACL) and PROVE-IT trials showed greater incidences (2.5% and 3.3%, respectively) in liver enzyme elevation, defined as 3 or more times the ULN.[47,50] In addition, simvastatin at 80-mg doses showed a slightly greater incidence (0.4%) in myopathy defined as CK more than 10 times the ULN with symptoms and incidence (0.9%) of liver enzymes 3 or more times the ULN in the A to Z trial.[48] Two ongoing trials should further help define the efficacy and safety of high-dose simvastatin therapy and its use in acute coronary syndrome. These studies include the SEARCH trial, a secondary prevention trial comparing simvastatin 80 mg and 20 mg, and the PROVE-IT trial in acute coronary syndrome, comparing simvastatin 40 mg alone with simvastatin 40 mg plus ezetimibe 10 mg in 10,000 patients.

Overall, the benefits of initiating early and higher-dose statins in patients with acute coronary syndrome seem to outweigh any of the potential adverse drug reactions that could occur. In addition, the initiation of drugs such as statins before hospital discharge not only can increase long-term adherence, but also can decrease reinfarction and mortality rates.[67,68]

CHRONIC KIDNEY DISEASE BACKGROUND

The population considered to have CKD according to the National Kidney Foundation's K/DOQI guidelines includes patients who have had at least 3 months of either structural or function abnormalities of the kidney or a glomerular filtration rate (GFR) of less than 60 mL/minute/1.73 m^2.[69] The National Kidney Foundation Task Force on Cardiovascular Disease has designated patients with CKD as having a higher incidence of atherosclerotic CVD (ACVD), which thus should be treated as a CHD risk equivalent.[70,71] This is primarily because these patients develop premature CVD, CHD, CVD, and/or peripheral vascular disease. There are differences between the prevalence of dyslipidemias treated with hemodialysis and peritoneal dialysis. A large cross-sectional analysis of adult patients on hemodialysis revealed that only 20% had normal lipid profiles (LDL less than 130 mg/dL, HDL more than 40 mg/dL, and triglycerides less than 150 mg/dL), and 61% were candidates for treatment; 55% specifically needed therapy for LDL more than 100 mg/dL.[72] Applying these same parameters to adult patients receiving peritoneal dialysis, only 15% had normal lipid profiles, and 78% were candidates for treatment; 73% needed therapy for LDL more than 100 mg/dL.[72] The prevalence of dyslipidemia is even greater in kidney transplant patients.[72] This evidence becomes the basis for the evaluation and assessment of dyslipidemia in this patient population.

SCREENING/EVALUATION

The K/DOQI guidelines recommend that all adults and adolescents with CKD be evaluated for dyslipidemia through a complete fasting lipid profile.[72] For all stable patients with stage 5 CKD, dyslipidemia should be evaluated on presentation, 2-3 months after any change in treatment or condition known to cause dyslipidemia, and at least annually thereafter. In addition, the assessment of lipids should occur before dialysis or on days when dialysis is not given. Finally, stage 5 patients should also be screened for common secondary causes of dyslipidemia such as nephrotic syndrome, hypothyroidism, excessive alcohol ingestion, and chronic liver disease. This is critical not only for preventing ACVD, but also for preventing decreased kidney function.[73] Although it is not known which condition worsens the other, it is important to target and treat both.

TREATMENT RECOMMENDATIONS

Assuming triglycerides are not greater than 500, the focus of management centers around achieving an LDL goal of less than 100 mg/dL in both adults and pediatric patients with CKD (see Table 6).[72] Although the K/DOQI guidelines address adolescents with dyslipidemia, the AHA more recently categorized pediatric patients with CKD as tier I (high risk) and made the LDL-C goal less than 100 mg/dL.[36] This new goal for pediatric patients is slightly more aggressive than the current K/DOQI guidelines, which do not recommend any treatment unless the LDL-C is more than 130 mg/dL, triglycerides are 200 mg/dL or more, and non-HDLs are 160 mg/dL or more.

TABLE 6. MANAGEMENT OF ADULTS WITH CKD AND DYSLIPIDEMIA

Lipid Level (mg/dL)	Goal (mg/dL)	Approaches to Treatment
TG ≥ 500	< 500	TLC → TLC + fibrate or niacin or omega-3 fatty acids
LDL 100–129	< 100	TLC → TLC + low-dose statin or ezetimibe → BAS or niacin
LDL ≥ 130	< 100	TLC + low-dose statin or ezetimibe → TLC + higher-dose statin ± ezetimibe → BAS or Niacin
TG ≥ 200 and HDL ≥ 130	Non-HDL < 130	TLC + low-dose statin or ezetimibe → TLC + higher-dose statin ± ezetimibe → BAS or niacin or omega-3 fatty acids

Note: Lipid levels and goals are from the 2003 K/DOQI guidelines for dyslipidemia.
BAS = bile acid sequestrant; CKD = chronic kidney disease; fish oil = omega-3 fatty acids refer to (EPA and DHA); TG = triglyceride; TLCs = therapeutic lifestyle changes.

As a result, any secondary causes, TLCs, statins, and fibrates (specifically gemfibrozil) are the mainstay of treatment for achieving these goals (see Table 6).[74-76] With the exception of rosuvastatin, most statins can be used safely in patients with CKD, but initial doses and maximal doses may need to be lowered (see Table 7). In fact, statins have been shown to slow the decline of kidney disease and reduce high-sensitivity CRP and may be safe and effective in patients undergoing hemodialysis.[77-84] The emergence of ezetimibe has also become a viable option that may help achieve LDL-C goals while using lower daily doses of statins without added drug interactions or increased risk of adverse effects. If the LDL-C remains 100 mg/dL or more despite statin therapy, another pharmacological option is the use of BABRs. However, they should not be used when triglycerides are 400 mg/dL or more and should be used cautiously when triglycerides are 200 mg/dL and above because they can further increase triglycerides.[72] In addition, their avoidance with concomitant drugs within a 2- to 4-hour period is often warranted, and GI adverse effects contribute to low compliance.

When triglycerides are either more than 500 mg/dL or more than 200 mg and non-HDL is more than 130 mg/dL, the use of fibrates, niacin, and fish oil may be beneficial.[72] Of the three options, the K/DOQI guidelines recommend fibrates with a focus primarily on gemfibrozil because it does not generally increase SCr concentrations and does not require dose adjustments in CKD like fenofibrate does.[85-88] Although the K/DOQI guidelines currently do not recommend a dosage adjustment of gemfibrozil, the Report of the National Lipid Association's Safety Task Force on Nonstatins recommends reductions in doses when the GFR is 15–59 mL/minute/1.73 m^2 or avoidance if less than 15 mL/minute/1.73 m^2 (see Table 7).[89] However, few data are available from the K/DOQI guidelines with regard to fish oils (e.g., omega-3 fatty acids; EPA and DHA); thus, they are generally recommended. However, several studies continue to suggest they are safe and effective in patients with CKD in doses up to 2.4 g (EPA+DHA) per day.[74-76] Their lack of drug interactions, costs, availability OTC, lack of renal dose adjustments, and efficacy in reducing triglycerides make them attractive options. However, no long-term data exist to determine their impact on cardiovascular outcomes in this population.

HUMAN IMMUNODEFICIENCY VIRUS

BACKGROUND

The development of metabolic complications, specifically dyslipidemia, insulin resistance, and lipodystrophy, in patients with HIV infection being treated with highly active antiretroviral therapy (HAART) can be a significant problem.[90-92] Although certain nucleoside reverse transcriptase inhibitor (NRTI)- and nonnucleoside reverse transcriptase inhibitor (NNRTI)-based HAART can cause these complications to various degrees, the PI-based HAART is currently known to be the worse offender.[93-103] With the exception of atazanavir (Reyataz; ATV), the frequency of PI-related dyslipidemia can be as high as 47–54%.[104] The development of dyslipidemia is multifactorial but is largely caused by PI-related insulin resistance. The overall estimated incidence of diabetes mellitus in patients with HIV infection on HAART is 3–5%. In patients with HIV infection and lipodystrophy treated with PI-based HAART, the

TABLE 7. RENAL DOSING CONSIDERATIONS FOR LIPID-LOWERING DRUGS

Treatment	GFR (mL/minute/1.73 m2) Stage 1 60–89	Stage 2 30–59	Stage 3 15–29	Stage 4 < 15 or Dialysis	Comments
HMG-CoA reductase inhibitors (mg)					
Atorvastatin	10–80	10–80	10–80	10–80	Substrate of CYP3A4
Fluvastatin	10–80	10–80	10–40	10–40	Doses > 40 mg not studied in severe renal impairment
Lovastatin	20–80	20–80	10–40	10–40	Package insert warns against use doses > 20 mg when CrCl < 30 mL/minute
Pravastatin	20–40	20–40	20–40	20–40	Package insert: if CrCl < 30 mL/minute starting dose is 10 mg
Rosuvastatin	5–40	5–40	5–10	?	Renal dosing in dialysis not known (Css 50% greater)
Simvastatin	20–80	20–80	10–40	10–40	Package insert: if CrCl < 30 mL/minute starting dose is 5 mg
Cholesterol absorption inhibitors (mg)					
Ezetimibe	10	10	10	10	In CrCl ≤ 30, increase mean AUC by 1.5-fold
Bile acid binding resins (g)					
Colesevelam	2.6–3.8				No dose adjustments; not absorbed
Cholestyramine	4–16				No dose adjustments; not absorbed
Colestipol	5–20				No dose adjustments; not absorbed

Treatment	GFR (mL/minute/1.73 m2) Stage 1 60–89	Stage 2 30–59	Stage 3 15–29	Stage 4 < 15 or Dialysis	Comments
Fibric acid derivatives (mg)					
Fenofibrate[a]	201	134	67	Avoid	Product inserts: recommended lowest starting dose impaired renal function and provide no dose limits doses reflect max doses per K/DOQI guidelines; Can also ↑ SCr
Gemfibrozil	1200	1200 [600/day]	1200 [600/day]	1200 [avoid]	This is total daily dose (600 mg twice daily) (recommendation by the NLA)
Niacin (g)	1–2 g	1–2 g	1–2 g	↓ to 50%	Product insert: recommends caution as not studied in CKD; doses listed are for extended- and sustained-release formulations
Omega-3 fatty acids (g)	1.5–2.4	1.5–2.4	1.5–2.4	1.5–2.4	Dosing ranges studied; reductions in TG of 4–52%
Plant stanol/sterols	2 g	2 g	2 g	2 g	Listed by K/DOQI as component in TLC for adults with CKD

Note: All doses are listed as milligram per day unless otherwise stated and are based on ADT III, K/DOQI recommendations, National Lipid Association, and manufacturer package insert recommendations (see comments to potential differences).

[a]Available in multiple brands with different milligram doses (TriCor, Lofibra tablets, Lofibra capsules, Antara).

AUC = area under the curve; CKD = chronic kidney disease; CrCl = creatinine clearance; Css = concentrations at steady state; CYP = cytochrome P450; GFR = glomerular filtration rate; HMG-CoA = 3-hydroxy-3-methylglutaryl coenzyme A; K/DOQI = Kidney Disease Outcomes Quality Initiative; NLA = National Lipid Association's 2007 Safety Task Force Report: The Nonstatins; SCr = serum creatinine; TG = triglyceride; TLC = therapeutic lifestyle changes.

prevalence of impaired glucose tolerance and type 2 diabetes mellitus is 35% and 7%, respectively.[105]

As a result, these patients have lipid profiles similar to patients with metabolic syndrome and/or type 2 diabetes mellitus (decrease in lipoprotein lipase; decrease in triglyceride clearance; shift to a smaller, more dense LDL; and increase in Lp(a)).[106-110] In addition, this patient population has been shown to have endothelial dysfunction. This translates into an increased risk of accelerated atherosclerosis and resulting CHD.[111-113]

The increase in CVD is supported by the Data Collection on Adverse Events of Anti-HIV Drugs Study Group, which is a large observational study consisting of about 23,400 patients from 11 cohorts in Europe, Australia, and the United States.[113] A recent analysis revealed that the adjusted relative risk (RR) of myocardial infarction per year of exposure to HAART was 1.16 (95% CI; 1.09–1.23), with most of the effect influenced by those taking PIs.[114] When patients taking PI-based HAART were controlled for lipids, the RR was 1.10 (p<0.05).[114] This increased risk despite controlling for lipids suggests that PIs are contributing to CVD through other mechanisms. When adjusting for lipids in patients taking NNRTI-based HAART, the RR was 1.0, further supporting the greater role of PIs in CVD.[114]

Now that patients infected with HIV taking HAART are living longer and their need for lifelong therapy may be increasing their risk of CVD, it is increasingly important to treat these metabolic complications proactively.

TREATMENT RECOMMENDATIONS

The approach to treating dyslipidemia in patients infected with HIV is not only multifactorial, but can also be plagued with increased adverse effects or adverse drug reactions, clinically significant drug–drug interactions, increased pill burden or fatigue, and increased financial strains. As a result, it can be difficult to get these patients to meet their cholesterol goals. Therefore, several interventions will likely be needed to improve their metabolic complications while also trying to avoid compromising their HIV viral or immunologic control. The Infectious Diseases Society of America and Adult AIDS Clinical Trials Group (IDSA/ACTG) proposed a set of guidelines for the management of patients with HIV infection with dyslipidemia.[92] In general, these guidelines reinforce many of the recommendations and goals promoted by the NCEP ATP III and further reinforce lifestyle modifications as a key component (see Table 8).[92,115] However, an intervention by the IDSA/ACTG that differs from the NCEP ATP III is the consideration of antiretroviral switch strategies without compromising viral load control (see Table 9).[92,97,99,116-128]

Antiretroviral switch strategies should only be tried in consultation with the treating HIV specialist. Although many patients may not be candidates for antiretroviral switching, the lipid-lowering benefits can be clinically significant and minimize the need for more aggressive drug therapy. Strategies known to improve both lipid profiles and/or insulin resistance include replacing the thymidine analogue NRTIs (zidovudine [Retrovir] and stavudine [Zerit]) with the non-thymidine analogue NRTIs, switching within NNRTIs (in particular efavirenz [Sustiva] to nevirapine

TABLE 8. APPROACHES TO THE EVALUATION AND MANAGEMENT OF HIV-RELATED DYSLIPIDEMIA

Step I:	Baseline lipid profile and within 3–6 mo of starting HAART
Step II:	Determine CHD risk factors and determine level of risk If ≥ 2 risk factors, perform 10-yr risk calculation.
Step III:	Treat secondary causes and modifiable risk factors (e.g., diet, smoking)
Step IV:	If not at NCEP III lipid goals despite TLC, consider: • Antiretroviral switch in appropriate candidates (only in consultation with HIV specialist/physician) and/or • Start lipid-lowering therapy (go to Step V)
Step V:	Evaluate TG concentrations • If TGs > 500: Initiate fibrate; alternative: niacin or omega-3 fatty acids • If TGs < 500 and LDL not at goal and TGs 200–500 + ↑ non-HDL:Initiate statin ± ezetimibe;[a] alternative: fibrate or niacin

Note: Based on Infectious Diseases Society of America/AIDS Clinical Trials Group Guidelines for the Evaluation and Management of HIV-Related Dyslipidemia.
[a]Not in current guidelines because of insufficient data at time of publication; added by author.
CHD = coronary heart disease; HAART = highly active antiretroviral therapy; LDL-C = low-density lipoprotein cholesterol; TLC = therapeutic lifestyle changes; TG = triglyceride.

TABLE 9. LIPID-LOWERING STRATEGIES OF SWITCHING ANTIRETROVIRALS IN PATIENTS INFECTED WITH HIV TAKING HAART

Antiretroviral Switch	TC (%)	TG (%)	HDL-C (%)	LDL-C (%)	Non-HDL (%)	Glucose
Switching within NNRTIs						
EFV to NVP	↓ 9–12	↓ 27	↑ 7–15	↓ 5–12	↓ 13–22	
NVP to EFV[a]	0	↑ 36	↓ 8	↑ 29	↑ 28	
Switching from PI to NNRTI						
PI to EFV	↓ 5.5	↓ 10	↑ 13	↓ 1.5	↓ 3.5	↑ 8 %
PI to NVP	↓ 11	↓ 26–32	↑ 15–20	↓ 19.5	↓ 9.5	↓ 3.5 %
Switching within PIs						
PI to ATV/RTV	↓ 10	↓ 37	↑ 4	↓ 7.5	↓ 15	↑ 2 mg/dL

Note: Should ONLY be considered in consultation with HIV specialist; percent changes are based on median or average change across studies; most studies have samples of < 100 subjects.
[a]Switch not recommended (presented to show the lipid effect can potentially be worse).
ATV = atazanavir (Reyataz); EFV = efavirenz (Sustiva); HDL-C = high-density lipoprotein; LDL-C = low-density lipoprotein; NNRTI = nonnucleoside reverse transcriptase inhibitor; NVP = nevirapine (Viramune); PI = protease inhibitor; RTV = ritonavir (Norvir); TC = total cholesterol; TG = triglyceride.

[Viramune]), switching from PI to NNRTI-based HAART, and switching from a non-ATV (Reyataz)-containing PI to a ritonavir (Norvir; RTV)-boosted ATV-based HAART (see Table 9).[92,97,99,116-128] Evidence from two retrospective studies suggests that there is an improvement in the lipid profile when switching from efavirenz to nevirapine but not when switching from nevirapine to efavirenz. This potential difference in lipid effect within the NNRTIs can also be seen when patients taking PI-based HAART are switched to an NNRTI-based HAART regimen.[116,117] Switching from a PI-based HAART to a nevirapine-based HAART appears to be more beneficial than switching to an efavirenz-based regimen.[118-123] When switching within the PIs to an ATV/RTV-based regimen, improvements have been made, not only in lipids but also in peripheral insulin sensitivity.[124-128]

If antiretroviral switching is not an option and/or additional lipid lowering is needed, pharmacological interventions can be safely and effectively implemented, but they require some very important considerations (see Table 10).[129-143]

Assuming that triglycerides are less than 500 mg, that underlying secondary causes (hypothyroidism, nephritic syndrome, liver disease, alcohol abuse, and uncontrolled diabetes) for dyslipidemia have been treated, and that the LDL-C is still above the recommended NCEP ATP III goal, statins can be a safe and effective option. However, caution should be used when choosing a particular statin and the starting doses, which are largely determined by the patient's current HAART regimen. The greatest concern for the safe use of statins is in patients taking PIs because all statins can inhibit one or more of the CYP isoenzymes (see Table 11).[104] With the exception of lovastatin and simvastatin, all of the currently available statins can be used in patients with HIV infection taking PI; however, most of the studies were performed with lower doses. Although rosuvastatin may not appear to interact with PIs because of its limited dependency on the CYP2C9 enzyme for metabolism, recent pharmacokinetic studies with the concomitant use of lopinavir/RTV showed a 2-fold increase in rosuvastatin levels.[144] Statin use in NNRTI-based HAART (specifically efavirenz-based HAART) would be less of a concern for all of them because they are known inducers of the CYP isoenzymes. Because of this induction, NNRTIs tend to reduce the areas under the curve of many statins, thereby potentially attenuating their lipid-lowering effects.[145] If additional LDL-C lowering is needed or if patients cannot tolerate statins at either high or low doses, then adding ezetimibe may be both a safe and effective alternative. Preliminary studies show that ezetimibe can lower LDL-C by 20% as monotherapy and by 7.5–25% when added to existing statin therapy without an increase in major adverse effects or loss of viral control.[141,142,143] However, no pharmacokinetic studies have been performed to date with ezetimibe coadministered with any of the antiretrovirals.

If the triglycerides are more than 500 mg/dL or more than 200 mg/dL and the patient is still above his or her non-HDL goal, then fibric acid derivatives and omega-3 fatty acids in dosages ranging from 1 to 6 g per day have been shown to be both safe and effective as monotherapy and in combination with each other and/or statins (see Table 10).[130-139] The increases seen in LDL-C with fibrate and omega-3 fatty acid use are largely because of changes in the density/buoyancy of LDL-C rather than particle count and should not be viewed as a worsening of the LDL. Omega-3 fatty acids can

TABLE 10. LIPID-LOWERING EFFICACY OF DRUG INTERVENTIONS IN PATIENTS INFECTED WITH HIV WITH DYSLIPIDEMIA

Drug Class	Doses Studied	TC (%)	TG (%)	HDL-C (%)	LDL-C (%)
Statins					
Atorvastatin	10 mg	↓ 21–46	↑ 3	↓ 37	
Fluvastatin	20–80 mg	↓ 17	↓ 0–5	↓ 0–8	↓ 24
Lovastatin		↓ 29	↓ 42		
Pravastatin	10–40 mg		↓ 13–41	↑ 0–10	↓ 14–39
Rosuvastatin	10 mg	↓ 30	↑ 28	↓ 22	
Simvastatin[a,b]	20–40 mg	↓ 20	↓ 20–29	0	↓ 36
Fibric acid derivatives					
Fenofibrate	200 mg	↓ 5–14	↓ 35–58	↑ 11–15	↑ 8–50
Gemfibrozil	600 mg BID	↓ 3–7	↓ 19–40	↑ 1–11	
Statin + fibrate					
Fenofibrate + statin	200 mg	↓ 16	↓ 45	↑ 17	↓ 5
Gemfibrozil + statin	600 mg BID	↓ 22	↓ 42–60	↑ 5	
Fish oil	1–6 g/day	0	↓ 12–46	↓ 1.7	↑ 22
Fish oil + fenofibrate	6 g + 160 mg/day		↓ 59	↑ 38–64	
Niacin	2 g/day	↓ 14	↓ 34	↑ 3	↓ 4
Cholesterol absorption inhibitor					
Ezetimibe	10 mg	↓ 10	0	0	↓ 20
Ezetimibe + statin	10 mg	↓ 5–11	↓ 2–22	↑ 8	↓ 7.5–25

Note: Blank areas reflect no known data, or data were not reported in studies.
[a] Should not be used in protease inhibitor–based regimens; is an option in nonnucleoside reverse transcriptase inhibitor–based high active antiretroviral therapy.
[b] Data primarily reflect efficacy in efavirenz-based regimens.
BID = 2 times/day; HDL-C = high-density lipoprotein cholesterol; LDL-C = low-density lipoprotein cholesterol; TC = total cholesterol; TG = triglyceride.

decrease inflammation and potentially inhibit the immune system; in this regard, one prospective study showed that there were no detrimental effects on the body's ability to maintain HIV viral control and/or replication.[138] Although niacin is recommended by the IDSA/ACTG guidelines as an alternative agent, the limited data in this population; the ability of niacin to worsen insulin resistance, a problem already exacerbated with PI use; and its tolerability profile make it a less desirable option.[92,140]

Finally, the development of the new class of antiretrovirals, the integrase inhibitors, may require additional consideration especially if the patient with HIV is given a statin. The only drug in this class so far is raltegravir (Isentress); it has been associated with increased CK concentrations of 2.2–2.4%, with some cases of rhabdomyolysis being reported in the pooled data of clinical trials.[146] Although no direct cause has

TABLE 11. STATIN USE IN PATIENTS INFECTED WITH HIV TAKING HAART

Characteristic	Lovastatin	Pravastatin	Simvastatin	Atorvastatin	Fluvastatin	Rosuvastatin
CYP isoenzyme	3A4 (major substrate)	Sulfation	3A4 (major substrate)	3A4 (mod substrate)	2C9 (75%); 3A4 (2.5%)	2C9 (< 10%); 2C19
NNRTIs						
Efavirenz	AUC ↓ 40%	AUC ↓ 58%	AUC ↓ 43%			
Nevirapine						
Delavirdine	Avoid	Avoid	Low dose			
Protease inhibitors						
Atazanavir	Avoid	Avoid	Low dose			
Fosamprenavir	Avoid	Avoid	AUC ↑150%			
Indinavir	Avoid	Avoid	Low dose			
Lopinavir/RTV	Avoid	AUC ↑ 33%	Avoid	AUC ↑ 5.8 x		AUC ↑ 5–8 x
Nelfinavir	Avoid	AUC ↑ 505%	AUC ↑ 74%			
Saquinavir/RTV	Avoid	Level ↓ 50%	AUC ↑ 3059%	Level ↑ 450%		
Tipranavir/RTV	Avoid	Avoid	AUC ↑ 9 x			
Darunavir/RTV	Avoid	AUC ↑ 5-fold	Avoid	Low dose		

AUC = area under the curve; CYP = cytochrome P450; HAART = highly active antiretroviral therapy; NNRTI = nonnucleoside reverse transcriptase inhibitor; RTV = ritonavir.

TABLE 12. EFFECT OF IMMUNOSUPPRESSIVE AGENTS ON CARDIOVASCULAR-RELATED RISK FACTORS IN PATIENTS WITH ORGAN TRANSPLANTATION

Variable	Azathioprine	Cyclosporine	Mycophenolate	Sirolimus	Tacrolimus
Metabolism	HGRT; then TPMT & XO	Sub/Inh CYP3A4, P-gp, & OAT1B1	Glucuronidation	Sub-CYP3A4, P-gp	Sub-CYP3A4, P-gp, & OATP1B1
Incidence in ↑ Lipids vs. placebo	28–38% vs. 23%		41% vs. NR	38–57% vs. 23%	31% vs. 38%
Incidence of IDDM					11–20%
Incidence of HTN vs. placebo	29% vs. 48%	8–26% vs. 2%	28–77% vs. NR	39–49% vs. 48%	38–47% vs. 56%
Incidence of ↑ Cr vs. placebo	28–36% vs. 38%	39–49% vs. 13%	39% vs. NR	35–40% vs. 38%	24–39% vs. 19–25%

Cr = creatinine; CYP = cytochrome P450; HGRT = hypoxanthine-guanine phosphoribosyltransferase; HTN = hypertension; IDDM = insulin-dependent diabetes mellitus; Inh = inhibitor; NR = not reported in package insert; OATP1B1 = organic anion transport polypeptide; P-gp = P-glycoprotein; SUB = substrate; TPMT = thiopurine S-methyltransferase; XO = xanthine oxidase.

been established, there are also no data evaluating an increased risk of myopathy or rhabdomyolysis in patients with HIV receiving raltegravir plus statins and/or fibrates. However, until further information is available, additional consideration is warranted in the overall management of the patient with HIV taking this drug.

SOLID ORGAN TRANSPLANTATION

BACKGROUND

Significant developments in immunosuppressive therapy during the past two decades have led to improvements in rejection-related mortality.[147] However, as many as 80% of heart transplant and 60% of renal transplant patients will develop posttransplant hyperlipidemia.[148,149] Several large-scale studies have demonstrated an increased risk of significant atherosclerosis that may also facilitate the development of graft coronary vasculopathy in heart transplant patients.[150,151] There is no set LDL threshold for the prevention or treatment of graft coronary vasculopathy. The worsening of lipids and other cardiovascular-related risk factors (hypertension and diabetes) are now known to be exacerbated, in part, by several of the immunosuppressive therapies (calcineurin inhibitors; see Table 12).[152-157] Elevations in triglycerides, LDL-C, and apolipoprotein B can occur as early as 3 months after therapy begins.[158,159] Although lipid elevations decline over time, they largely remain elevated with respect to pretransplant levels.

TREATMENT RECOMMENDATIONS

Because of the adverse effects of immunosuppressive drugs on a transplant recipient's lipid profile, consideration should be given to altering the immunosuppression regimen, especially if maximal medical management has failed.[72,147] This should only be done in direct consultation with the transplant physician overseeing the posttransplant care of the patient, with significant consideration given to the risk-to-benefit ratio.[72] With the type of organ transplantation in mind, several strategies have been studied, including tapering and discontinuing prednisone with or without adding or adjusting the dose of other immunosuppressive agents (such as azathioprine or mycophenolate mofetil); tapering and discontinuing cyclosporine with or without adding or adjusting the dose of other immunosuppressive agents (such as azathioprine or mycophenolate mofetil); switching cyclosporine for tacrolimus; and stopping or replacing sirolimus with another immunosuppressive agent.[160-171] Most of these strategies have been tried in patients with kidney transplants.

When lipid-lowering therapy is needed, statins provide the greatest degree of LDL-C lowering; however, they may require dose adjustments. Most of the drug interactions surrounding the use of statins largely exist with the concomitant use of cyclosporine and thus require dose readjustments and additional monitoring considerations (see Table 13).[7-13,17,72,154,172-177] Although most of the pharmacokinetic data exist with cyclosporine, a recent pharmacokinetic study evaluating the effects of tacrolimus on atorvastatin levels showed no interactions.[178] This is probably because tacrolimus is only a substrate for CYP3A4 and not an inhibitor, as seen with cyclosporine. Although it may appear that rosuvastatin would not interact with these immunosuppressives, cyclosporine has been shown to increase the levels of rosuvastatin 4-fold.[11] This is most likely mediated through cyclosporine's inhibition of organic anion transporting polypeptide (OATP).

If the statins either cannot be tolerated or do not allow the patient to achieve his or her LDL-C goal, ezetimibe may be used; its use has been increasing in this population. To date, most of the safety and efficacy data with ezetimibe exist within kidney transplant patients. The LDL-C lowering achieved among one prospective study and two retrospective studies in which ezetimibe was added to existing statin therapy ranged from 20% to 35%.[179-181] Although the sample sizes for these studies were all less than 35 patients, these initial data support beneficial effects and safety. Of interest, the LDL-C lowering is greater than that generally observed (about 20% additional lowering when added to a statin) in nontransplant patients.[13] It is not known if the increased levels of ezetimibe in patients taking concomitant cyclosporine resulted in this potential difference. Other options for lowering LDL-C include BABRs, for which there is very little evidence in either kidney or heart transplant patients. In addition to the adverse effects related to BABRs, the potential interaction with immunosuppressive drugs and less effective LDL-C lowering compared with other options have made the use of these drugs less desirable.[182]

When triglycerides are the primary problem or become a secondary target for achieving non-HDL goals, the use of fibric acid derivatives should probably be limited to gemfibrozil.[72] Although both fenofibrate and gemfibrozil are effective in reducing triglycerides in organ transplant patients, there were significantly more dropout rates (67%) and adverse drug reactions (elevations in SCr, elevations in CK, and acute

TABLE 13. KNOWN EFFECTS OF CYCLOSPORINE ON LIPID-LOWERING DRUGS IN PATIENTS WITH ORGAN TRANSPLANTATION

Drug	↑ in AUC of Lipid	Comment
Atorvastatin	8-fold	K/DOQI recommends daily doses of 10–40 mg; PI = NR
Fluvastatin	2-fold	K/DOQI recommends daily doses of 10–40 mg; PI = NR
Lovastatin	2- to 20-fold	K/DOQI recommends daily doses of 10–40 mg; PI = 10-20 mg
Pravastatin	5-fold	K/DOQI recommends daily doses of 20–40 mg; PI = precaution
Rosuvastatin	7-fold	K/DOQI NR; PI = max dose is 5 mg
Simvastatin	3- to 8-fold	K/DOQI recommends daily doses of 10–40 mg; PI = 10 mg max
Ezetimibe	3.4-fold	K/DOQI NR; PI = caution with concomitant use and monitor cyclosporine levels because they can also be increased
Fenofibrate		Caution warranted because both drugs can ↑ SCr

AUC = area under the curve; K/DOQI = Kidney Disease Outcomes Quality Initiative; NR = no recommendation made; PI = package insert (recommendations).

rejection episodes) in a small study of cardiac transplant patients on fenofibrate.[183] Several studies have shown that fenofibrate not only accumulates in patients with compromised renal function, but also can increase SCr concentrations.[85-88] However, this increase in SCr does not appear to worsen GFRs in nontransplant patients.[89] Several studies representing both kidney and heart transplant patients have also shown that gemfibrozil is not only efficacious but also well tolerated.[158,184,185] However, additional monitoring for adverse drug reactions is warranted if gemfibrozil is used in combination with statin therapy, and dose reductions may also be warranted because of accumulations in severe renal impairment.[186,187] Another option gaining interest is omega-3 fatty acids. Preliminary studies in heart transplantation have suggested that endothelium-dependent coronary vasodilation improved after 3 weeks of 5 g per day of omega-3 fatty acids, which may be contributing to the inhibition of graft coronary vasculopathy and longer cardiac graft survival observed in animal cardiac transplant models.[188,189] However, the overall safety and efficacy and influence on survival in organ transplantation are not known. Finally, niacin has limited studies as well; however, although it is effective, it is not generally tolerated.[190,191]

ELDERLY

BACKGROUND

The absolute risk of CHD mortality and risk attributable to elevated cholesterol increases dramatically with age.[192,193] The time in which these cardiovascular events occur is also dramatically shorter, and the absolute risk of increased cholesterol becomes greater with age. Patients older than 65 years are not only at greater risk of sudden death, but also account for more than 60% of hospital admissions for acute myocardial infarction and have higher rates of both in-hospital and postdischarge death than patients younger than 65 years.[194-196] These findings suggest that the greatest impact on public health is risk factor management in the elderly.

The growing level of evidence supports that the 1%/2% rule (the concept that a 1% lowering of cholesterol reduces CHD by about 2% in middle-aged people) is also true in high-risk elderly patients.[193] This evidence is partly based on several recent clinical trials that included older patients.[197-204] A recent meta-analysis of data, which represented patients 65 years and older from the LIPID, 4S, CARE, and AFCAPS/TexCAPS trials, showed a 32% reduction in RR factors for a major coronary event, a benefit similar to patients younger than 65 years.[205] In fact, the absolute risk reduction was greater for those older than 65 versus younger than 65: 44 events per 1000 patients versus 32 events per 1000 patients, respectively. This translates into a number needed to treat of 23 for patients older than 65 and 31 for patients younger than 65 years. The benefit of statins in the elderly was further substantiated by a recent meta-analysis using nine clinical trials representing 19,569 patients ranging from 65 to 82 years.[206] The investigators found a 22% RR reduction during 5 years, a reduction in CHD mortality by 30%, in nonfatal myocardial infarction of 26%, in need for revascularization of 26%, and in stroke by 25%. They also estimated that the number needed to treat to save 1 life was 28. In addition, recent data support that statin therapy is associated with better long-term mortality in elderly patients hospitalized with congestive heart

TABLE 14. CLINICAL TRIALS OF LIPID-LOWERING DRUGS IN THE ELDERLY

Trial	Total (n)/ Elderly (n)	Treatments	Results
HPS	20,536/5806 > 70 yr	Simvastatin 40 mg vs. placebo	20% fewer vascular events and was well tolerated and safe
LIPID	9014/3514 > 65 yr	Pravastatin 40 mg vs. placebo	Reductions of 21% in all-cause mortality, 24% in CHD death, 26% in MI, and 12% in stroke
4S	4444/1021 > 65 yr	Simvastatin 20–40 mg vs. placebo	Reductions of 34% in all-cause mortality, 43% in CHD mortality, 34% in major coronary events, and 41% in revascularizations
CARE	4159/1283 > 65 yr	Pravastatin 40 mg vs. placebo	32% RR in major coronary events, 45% RR in CHD mortality, 40% RR in stroke
AFCAPS/ TexCAPS	6605/1416 > 65 yr	Lovastatin 20–40 mg vs. placebo	Lovastatin attenuated the risk conferred by sex, age, family history, HTN, smoking, LDL level, and low HDLs.
PROSPER	5807/5807 ages 70–82	Pravastatin 40 mg vs. placebo	Reductions of 19% in CHD mortality without effect on cognitive function or disability
SAGE	893 ages 65–85	Atorvastatin 80 mg vs. pravastatin 40 mg	Reduction of 67% in all-cause mortality (p=0.014), 29% in major cardiovascular events (HR, 0.71; 95% CI, 0.46–1.09; p=0.114) and in duration of ischemia
VA-HIT	2531/1265 > 66 yr	Gemfibrozil 1200 mg vs. placebo	26% CHD RR (coronary death, nonfatal MI, confirmed stroke); only men studied

AFCAPS/TexCAPS = Air Force Coronary Atherosclerosis Prevention Study; CARE = Cholesterol and Recurrent Events; CHD = coronary heart disease; 4S = Scandinavian Simvastatin Survival Study; HPS = Heart Protection Study; HTN = hypertension; LIPID = Long-term Intervention with Pravastatin in Ischemic Disease; MI = myocardial infarction; PROSPER = Prospective Study of Pravastatin in the Elderly at Risk; RR = risk reduction; RRR = relative risk reduction; SAGE = Study Assessing Goals in the Elderly; VA-HIT = Veterans Affairs High-Density Lipoprotein Cholesterol Intervention Trial.

failure.[207] Despite these supportive data, few elderly patients receive appropriate lipid-lowering treatment, even after a cardiovascular event.[208,209] There are several reasons for this such as cost, increased risk of drug interactions, and greater incidence of adverse effects. In addition, there have been concerns that statins may worsen cognitive function and potentially increase the risk of cancer. However, neither has occurred; statins may, in fact, help prevent the decline of cognitive impairment.[210–213]

TREATMENT RECOMMENDATIONS

The implementation of age-specific and appropriate diet and exercise programs is still of major importance in the elderly and should not be minimized. When drug therapy is needed, factors such as greater risk of drug–drug interactions, adverse effects, decreased elimination, and cost can be prohibitive to achieving adequate cholesterol lowering. The greatest amount of evidence in reducing cardiovascular mortality lies with the use of statins (see Table 14). However, adverse effects and complications with some drugs can also be potentially more problematic in the elderly. For example, the occurrence of gallstones may be increased with the use of fibrates, with cholecystectomy procedures carrying a greater risk in this population. Drugs such as BABRs may significantly interact with other concomitant drugs and cause unwanted GI adverse effects. Dose reductions may be necessary for fenofibrate because of decreased renal function; thus, lower initial doses may be warranted. So far, ezetimibe has been shown to be a safe and effective treatment option in the elderly. In a recent pooled analysis of four clinical trials, ezetimibe was equally tolerable and effective on LDL-C, triglycerides, and HDL-C and was independent of age group (age younger than 65 vs. 65 years or older; age younger than 75 and 75 years and older).[214] Despite these potential setbacks to safe and effective use, the greatest problem facing this population is their underuse. The Clinical Quality Improvement Network Investigators found that among 3304 hospitalized patients at high risk of future CVD events, only 5% of older patients had been prescribed lipid-lowering therapy.[215] Data remain consistent with the decreased use of lipid-lowering therapy in this population.[208,209] More recently, the efficacy of high-dose atorvastatin in patients older than 65 with acute coronary syndrome was further described in a post hoc analysis of the MIRACL trial.[216] Finally, statin use in elderly patients hospitalized for heart failure may be associated with better long-term mortality, further supporting statin use in this population.[217]

Current evidence clearly supports the risk of CVD in this patient population and the benefits of appropriately prescribed and monitored drug therapy. It appears that elderly patients with established atherosclerotic disease or high-risk elderly patients without atherosclerotic disease should be considered for aggressive lipid lowering. In conclusion, excluding the elderly from lipid-lowering therapy because of age alone is both inappropriate and not recommended.[115,193]

REFERENCES

1. Amundsen AL, Khoury J, Iversen PO, et al. Marked changes in plasma lipids and lipoproteins during pregnancy in women with familial hypercholesterolemia. Atherosclerosis 2006;189:451-7.
2. Salameh WA, Mastrogiannis DS. Maternal hyperlipidemia in pregnancy. Clin Obstet Gynecol 1994;37:66-77.
3. Floridia M, Guaraldi G, Tamburrini E, et al. Lipodystrophy is an independent predictor of hypertriglyceridemia during pregnancy in HIV-infected women. AIDS 2006;20:944-7.
4. Sattar N, Greer IA, Louden J, et al. Lipoprotein subfraction changes in normal pregnancy: threshold effect of plasma triglyceride on appearance of small, dense low-density lipoprotein. J Clin Endocrinol Metab 1997;82:2483-91.
5. American Academy of Pediatrics. The transfer of drugs and other chemicals into human milk. Pediatrics 2001;108:776-89.
6. Hale TW. Medications and Mothers' Milk, 11th ed. Amarillo, TX: Pharmasoft Publishing, 2004:9.
7. Atorvastatin (Lipitor) [package insert]. New York: Pfizer; 2006.
8. Fluvastatin sodium (Lescol XL) [package insert]. East Hanover, NJ: Novartis Pharmaceutical; 2006.
9. Lovastatin (Mevacor) [package insert]. Whitehouse Station, NJ: Merck & Co.; 2005.
10. Pravastatin (Pravachol) [package insert]. Princeton, NJ: Bristol-Myers Squibb; 2002.
11. Rosuvastatin (Crestor) [package insert]. Wilmington, DE: AstraZeneca Pharmaceuticals; 2005.
12. Simvastatin (Zocor) [package insert]. Whitehouse Station, NJ: Merck & Co.; 2005.
13. Ezetimibe (Zetia) [package insert]. North Wales, PA: Merck/Schering-Plough Pharmaceuticals; 2006.
14. Colesevelam hydrochloride (Welchol) [package insert]. Parsippany, NJ: Daiichi Sankyo; 2006.
15. Cholestyramine (Questran) [package insert]. Princeton, NJ: Bristol-Myers Squibb; 1997.
16. Colestipol (Colestid) [package insert]. Kalamazoo, MI: Pharmacia & Upjohn; 1999.
17. Fenofibrate (TriCor) [package insert]. North Chicago, IL: Abbott Laboratories; 1998.
18. Gemfibrozil (Lopid) [package insert]. New York: Pfizer; 2002.
19. Niacin (Niaspan) [package insert]. Cranbury, NJ: Kos Pharmaceuticals; 2005.
20. Omega-3 acid ethyl esters 90 (Lovaza) [package insert]. Liberty Corner, NJ: Reliant Pharmaceuticals; June 2007.
21. Makino H, Harada-Shiba M. Long-term effect of low-density lipoprotein apheresis in patients with homozygous familial hypercholesterolemia. Ther Apher Dial 2003;7:397-401.

22. Manson JM, Freyssinges C, Ducrocq MB, Stephenson WP. Postmarketing surveillance of lovastatin and simvastatin exposure during pregnancy. Reprod Toxicol 1996;10:439-46.
23. Freyssinges C, Ducrocq MB. Simvastatin and pregnancy. Therapie 1996;51:537-42.
24. Pollack PS, Shields KE, Burnett DM, et al. Pregnancy outcomes after maternal exposure to simvastatin and lovastatin. Birth Defects Res A Clin Mol Teratol 2005;73:888-96.
25. Edison RJ, Muenke M. Central nervous system and limb anomalies in case reports of first-trimester statin exposure. N Engl J Med 2004;350:1579-82.
26. Garg A, Grundy SM. Cholestyramine therapy for dyslipidemia in non-insulin-dependent diabetes mellitus. A short-term, double-blind, crossover trial. Ann Intern Med 1994;121:416-22.
27. Glueck CJ, Tsang RC, Fallat RW, Scheel BA. Plasma vitamin A and E levels in children with familial type II hyperlipoproteinemia during therapy with diet and cholestyramine resin. Pediatrics 1974;54:51-9.
28. West RJ, Lloyd JK. The effect of cholestyramine on intestinal absorption. Gut 1975;66:92-98.
29. Watkins DW, Khalafi R, Cassidy MM, Vahouny GV. Alterations in calcium, magnesium, and zinc metabolism by dietary cholestyramine. Dig Dis Sci 1985:30;477-82.
30. Yen TH, Chang CT, Wu MS, Huang CC. Acute rhabdomyolysis after gemfibrozil therapy in a pregnant patient complicated with acute pancreatitis and hypertriglyceridemia while receiving continuous veno-venous hemofiltration therapy. Ren Fail 2003;25:139-43.
31. Blanchard DS. Omega-3 fatty acid supplementation in perinatal settings. MSC Am J Matern Child Nurs 2006;31:250-6.
32. Bays HE. Safety considerations with omega-3 fatty acid therapy. Am J Cardiol 2007;99(suppl):35C-43C.
33. Berger A, Jones P, Abumweis S. Plant stanols: factors affecting their efficacy and safety as functional food ingredients. Lipids Health Dis 2004;3:5.
34. de Jongh S, Lilien MR, Bakker HD, et al. Family history of cardiovascular events and endothelial dysfunction in children with familial hypercholesterolemia. Atherosclerosis 2002;163:193-7.
35. Ose L. Diagnostic, clinical, and therapeutic aspects of familial hypercholesterolemia in children. Semin Vasc Med 2004;4:51-7.
36. Kavey RW, Allada V, Daniels SR, et al. American Heart Association Scientific Statement. Cardiovascular risk reduction in high-risk pediatric patients. Circulation 2006;114:2710-38.
37. McCrindle BW, Urbina EM, Dennison BA, et al. Drug therapy of high risk lipid abnormalities in children and adolescents. A scientific statement from the American Heart Association Atherosclerosis, Hypertension, and Obesity in Youth Committee, Council of Cardiovascular Disease in the Young, with the Council on Cardiovascular Nursing. Circulation 2007;115:1948-67.

38. National Cholesterol Education Program. Report of the Expert Panel on Blood Cholesterol Levels in Children and Adolescents. Pediatrics 1992;89(pt 2):525–84.
39. Liacouras CA, Coates PM, Gallagher PR. Use of cholestyramine in the treatment of children with familial combined hyperlipidemia. J Pediatr 1993;122:477–82.
40. West RJ, Lloyd JK. Long-term follow-up of children with familial hypercholesterolemia treated with cholestyramine. Lancet 1980;2:873–5.
41. Rodenburg J, Vissers MN, Wiegman A, et al. Familial hyperlipidemia in children. Curr Opin Lipidol 2004;15:405–11.
42. Becker M, Staab D, Von Bergmann K. Treatment of severe familial hypercholesterolemia in childhood with sitosterol and sitostanol. J Pediatr 1993;122:292–6.
43. Gylling H, Siimes MA, Miettinen TA. Sitostanol ester margarine in dietary treatment of children with familial hypercholesterolemia. J Lipid Res 1995;36:1807–12.
44. Tammi A, Ronnemaa T, Gylling H, et al. Plant stanol ester margarine lowers serum total and low-density lipoprotein cholesterol concentrations of healthy children: the STRIP project. Special Turku Coronary Risk Factors Intervention Project. J Pediatr 200;136:503–10.
45. American Heart Association. Part 8: Stabilization of the patient with acute coronary syndromes. Circulation 2005;112:IV-89–IV-110.
46. Arntz HR, Agrawal R, Wunderlich W, et al. Beneficial effects of pravastatin (±/-cholestyramine/niacin) initiated immediately after a coronary event (the randomized Lipid-Coronary Artery Disease [L-CAD] Study). Am J Cardiol 2000;86:1293–8.
47. Cannon CP, Braunwald E, McCabe CH, et al. Intensive versus moderate lipid lowering with statins after acute coronary syndromes. N Engl J Med 2004;350:1495–504.
48. de Lemos JA, Blazing MA, Wiviott SD, et al. Early intensive vs a delayed conservative simvastatin strategy in patients with acute coronary syndromes: phase Z of the A to Z trial. JAMA 2004;292:1307–16.
49. Dupuis J, Tardif JC, Cernacek P, Theroux P. Cholesterol reduction rapidly improves endothelial function after acute coronary syndromes. The RECIFE (reduction of cholesterol in ischemia and function of the endothelium) trial. Circulation 1999;99:3227–33.
50. Kinlay S, Schwartz GG, Olsson AG, et al. High-dose atorvastatin enhances the decline in inflammatory markers in patients with acute coronary syndromes in the MIRACL study. Circulation 2003;108:1560–6.
51. Pedersen TR, Jahnsen KE, Vatn S, et al. Benefits of early lipid-lowering intervention in high-risk patients: the lipid intervention strategies for coronary patients study. Clin Ther 2000;22:949–60.
52. Schwartz GG, Olsson AG, Ezekowitz MD, et al. Effects of atorvastatin on early recurrent ischemic events in acute coronary syndromes: the MIRACL study: a randomized controlled trial. JAMA 2001;285:1711–8.

53. Kayikcioglu M, Can L, Evrengul H, Payzin S, Kultursay H. The effect of statin therapy on ventricular late potentials in acute myocardial infarction. Int J Cardiol 2003;90:63–72.
54. Kayikcioglu M, Can L, Kultursay H, Payzin S, Turkoglu C. Early use of pravastatin in patients with acute myocardial infarction undergoing coronary angioplasty. Acta Cardiol 2002;57:295–302.
55. Waters DD, Schwartz GG, Olsson AG, et al. Effects of atorvastatin on stroke in patients with unstable angina or non–Q-wave myocardial infarction: a Myocardial Ischemia Reduction with Aggressive Cholesterol Lowering (MIRACL) substudy. Circulation 2002;106:1690–5.
56. Chan AW, Bhatt DL, Chew DP, et al. Relation of inflammation and benefit of statins after percutaneous coronary interventions. Circulation 2003;107:1750–6.
57. Den Hartog FR, Van Kalmthout PM, Van Loenhout TT, et al. Pravastatin in acute ischaemic syndromes: results of a randomised placebo-controlled trial. Int J Clin Pract 2001;55:300–4.
58. Heeschen C, Hamm CW, Laufs U, et al. Withdrawal of statins increases event rates in patients with acute coronary syndromes. Circulation 2002;105:1446–52.
59. Spencer FA, Fonarow GC, Frederick PD, et al. Early withdrawal of statin therapy in patients with non–ST-segment elevation myocardial infarction: National registry of myocardial infarction. Arch Intern Med 2004;164:2162–8.
60. Fonarow GC, Wright RS, Spencer FA, et al. Effect of statin use within the first 24 hours of admission for acute myocardial infarction on early morbidity and mortality. Am J Cardiol 2005;96:611–6.
61. Bavry AA, Kumbhani DJ, Mood G, et al. Benefit of early statin therapy during acute coronary syndromes: a meta-analysis [O-9]. Society for Cardiovascular Angiography and Interventions (SCAI) 29th Annual Scientific Sessions, May 10–13, 2006, Chicago, IL.
62. Libby P, Aikawa M. Mechanisms of plaque stabilization with statins. Am J Cardiol 2003;91:4B–8B.
63. Libby P, Aikawa M. Effects of statins in reducing thrombotic risk and modulating plaque vulnerability. Clin Cardiol 2003;26:I11–I14.
64. Wiviott SD, de Lemos JA, Cannon CP, Blazing M et al. A tale of two trials. A comparison of the post-acute coronary syndrome lipid lowering trials A to Z and PROVE-IT-TIMI 22. Circulation 2006;113:1406–14.
65. Nissen SE, Tuzcu EM, Schoenhagen P, et al. Statin therapy, LDL cholesterol, C-reactive protein, and coronary artery disease. N Engl J Med 2005;352:29–38.
66. Ridker PM, Cannon CP, Morrow D, et al. C-reactive protein levels and outcomes after statin therapy. N Engl J Med 2005;352:20–8.
67. Fonarow GC, Gawlinski A, Moughrabi S, Tillisch JH. Improved treatment of coronary heart disease by implementation of a cardiac hospitalization atherosclerosis management program (CHAMP). Am J Cardiol 2001;87:819–22.
68. Peterson ED, Roe MT, Mulgund J, et al. Association between hospital process performance and out comes among patients with acute coronary syndromes. JAMA 2006;295:1912–20.

69. National Kidney Foundation. K/DOQI clinical practice guidelines for chronic kidney disease: evaluation, classification, and stratification. Am J Kidney Dis 2002;39:S1–S266.
70. Levey AS, Beto JA, Coronado BE, et al. Controlling the epidemic of cardiovascular disease in chronic renal disease: What do we know? What do we need to learn? Where do we go from here? National Kidney Foundation Task Force on Cardiovascular Disease. Am J Kidney Dis 1998;32:853–906.
71. National Kidney Foundation. K/DOQI clinical practice guidelines for managing dyslipidemias in chronic kidney disease. Am J Kidney Dis 2003;41(suppl 3):S1–S91.
72. National Kidney Foundation. K/DOQI clinical practice guidelines for managing dyslipidemias in chronic kidney disease. Am J Kidney Dis 2003;41(suppl. 3):S22–S38.
73. Fried LF, Orchard TJ, Kasiske BL. The effect of lipid reduction on renal disease progression: a meta-analysis. Kidney Int 2001;59:260–9.
74. Ando M, Sanaka T, Nihei H. Eicosapentanoic acid reduces plasma levels of remnant lipoproteins and prevents in vivo peroxidation of LDL in dialysis patients. J Am Soc Nephrol 1999;19:2177–84.
75. Svensson M, Christensen JH, Solling J, Schmidt EB. The effect of n-3 fatty acids on plasma lipids and lipoproteins and blood pressure in patients with CRF. Am J Kidney Dis 2004;44:77–83.
76. Fiedler R, Mall M, Wand C, Osten B. Short-term administration of omega-3 fatty acids in hemodialysis patients with balanced lipid metabolism. J Ren Nutr 2005;15:253–6.
77. Fried LF, Orchard TJ, Kasiske BL. Effects of lipid reduction on the progression of renal disease: a meta-analysis. Kidney Int 2001;59:260–9.
78. Bianchi S, Bigazzi R, Caiazza A, et al. A controlled, prospective study of the effects of atorvastatin on proteinuria and progression of kidney disease. Am J Kidney Dis 2003;41:565–70.
79. Kano K, Nishikura K, Yamada Y, et al. Effect of fluvastatin and dipyridamole on proteinuria and renal function in childhood IgA nephropathy with mild histological findings and moderate proteinuria. Clin Nephrol 2003;60:85–9.
80. Athros VG, Mikhaildis DP, Papageorgiou AA, et al. The effect of statins versus untreated dyslipidemia on renal function in patients with coronary heart disease. A subgroup analysis of the Greek atorvastatin and coronary heart disease evaluation (GREACE) study. J Clin Pathol 2004;57:728–34.
81. Vidt DG, Cressman HD, Harris S, et al. Rosuvastatin-induced arrest in progression of renal disease. Cardiology 2004;102:52–60.
82. Verma A, Ranganna KM, Reddy RS, et al. Effect of rosuvastatin on C-reactive protein and renal function in patients with chronic kidney disease. Am J Cardiol 2005;96:1290–2.
83. Wanner C, Krane V, Marz W, et al. Atorvastatin in patients with type 2 diabetes mellitus undergoing hemodialysis. N Engl J Med 2005;353:238–48.

84. Mason NA, Bailie GR, Satayathum S, et al. HMG-coenzyme a reductase inhibitor use in associated with mortality reduction in hemodialysis patients. Am J Kidney Dis 2005;45:119–26.
85. Broeders N, Knoop C, Antoine M, Tielemans C, Abramowicz D. Fibrate-induced increase in blood urea and creatinine: Is gemfibrozil the only innocuous agent? Nephrol Dial Transplant 2000;15:1993–9.
86. Devuyst O, Goffin E, Pirson Y, van Ypersele de Strihou C. Creatinine rise after fibrate therapy in renal graft recipients. Lancet 1993;341:840.
87. Hottelart C, el Esper N, Achard JM, et al. Fenofibrate increases blood creatinine, but does not change the glomerular filtration rate in patients with mild renal insufficiency. Nephrologie 1999;20:41–4.
88. Lipscombe J, Lewis GF, Cattran D, Bargman JM. Deterioration in renal function associated with fibrate therapy. Clin Nephrol 2001;55:39–44.
89. Davidson MH, Armani A, McKenney JM, et al. Safety considerations with fibrate therapy. Am J Cardiol 2007;99(suppl):3C–18C.
90. Carr A, Samaras K, Thorisdottir A, et al. Diagnosis, prediction, and natural course of HIV-1 protease-inhibitor-associated lipodystrophy, hyperlipidemia, and diabetes mellitus: a cohort study. Lancet 1999;353:2093–9.
91. Vigouroux C, Gharakhanian S, Salhi Y, et al. Diabetes, insulin resistance and dyslipidaemia in lipodystrophic HIV-infected patients on highly active antiretroviral therapy (HAART). Diabetes Metab 1999;25:225–32.
92. Dube M, Stein J, Aberg J, et al. Guidelines for the evaluation and management of dyslipidemia in human immunodeficiency virus (HIV)-infected adults receiving antiretroviral therapy: recommendations of the HIV Medicine Association of the Infectious Disease Society of America and the Adult AIDS Clinical Trials Group. Clin Infect Dis 2003;37:613–27.
93. Tal A, Dall L. Didanosine-induced hypertriglyceridemia. Am J Med 1993;95:247.
94. Saint-Marc T, Partisani M, Poizot-Martin I, et al. A syndrome of peripheral fat wasting (lipodystrophy) in patients receiving long-term nucleoside analogue therapy. AIDS 1999;13:1659–67.
95. Modest G, Fuller J. Abacavir and diabetes. N Engl J Med 2001;344:142–4.
96. Moyle GJ, Baldwin C. Lipid elevations during non-nucleoside RTI (NNRTI) therapy: a cross-sectional analysis. Antivir Ther 1999;4(suppl 2):54.
97. Matthews G, Moyle G, Mandalia S, et al. Absence of association between individual thymidine analogues or nonnucleoside analogues and lipid abnormalities in HIV-1-infected persons on initial therapy. J Acquir Immune Defic Syndr 2000;24:310–5.
98. Gallant JE, Staszewski S, Pozniak AL, et al. Efficacy and safety of tenofovir DF vs stavudine in combination therapy in antiretroviral-naive patients: a 3-year randomized trial. JAMA 2004;292:191–201.
99. Dube M, Parker P, Tebas P, et al. Glucose metabolism, lipid, and body fat changes in antiretroviral-naive subjects randomized to nelfinavir or efavirenz plus dual nucleosides. AIDS 2005;19:1807–18.

100. Manfredi R, Calza L, Chiodo F. An extremely different dysmetabolic profile between the two available nonnucleoside reverse transcriptase inhibitors: efavirenz and nevirapine. J Acquir Immune Defic Syndr 2005;38:236–8.
101. Carr A, Samaras K, Burton S, et al. A syndrome of peripheral lipodystrophy, hyperlipidemia and insulin resistance due to HIV protease inhibitors. AIDS 1998;12:F51–8.
102. Echevarria KL, Hardin TC, Weiner M, Graybill JR. Serum lipids in patients infected with human immunodeficiency virus on protease inhibitor antiretrovirals. J Appl Ther Res 1999;2:171–6.
103. Mulligan K, Grunfeld C, Tai VW, et al. Hyperlipidemia and insulin resistance are induced by protease inhibitors independent of changes in body composition in patients with HIV infection. J Acquir Immune Defic Syndr 2000;23:35–43.
104. DHHS Panel on Antiretroviral Guidelines for Adults and Adolescents—a working group of the Office of AIDS Research Advisory Council (ORAC). Guidelines for the use of antiretroviral agents in HIV-1-infected adults and adolescents (version 10/10/2006). Available at http://aidsinfo.nih.gov/contentfiles/AdultandAdolescentGL.pdf. Accessed July 31, 2008.
105. Hadigan C, Meigs JB, Corcoran C, et al. Metabolic abnormalities and cardiovascular disease risk factors in adults with human immunodeficiency virus infection and lipodystrophy. Clin Infect Dis 2001;32:130–9.
106. Grunfeld C, Pang M, Doerrler W, et al. Lipids, lipoproteins, triglyceride clearance, and cytokines in human immunodeficiency virus infection and the acquired immunodeficiency syndrome. J Clin Endocrinol Metab 1992;74:1045–52.
107. Feingold KR, Krauss RM, Pang M, et al. The hypertriglyceridemia of acquired immunodeficiency syndrome is associated with an increased prevalence of low-density lipoprotein subclass pattern B. J Clin Endocrinol Metab 1993;76:1423–7.
108. Periard D, Telenti A, Sudre P, et al. Atherogenic dyslipidemia in HIV-infected individuals treated with protease inhibitors. Circulation 1999;100:700–5.
109. Constans J, Pellegrin JL, Peuchant E, et al. High plasma lipoprotein (a) in HIV-positive patients. Lancet 1993;341:1099–100.
110. Stein J, Klein M, Bellehumeur J, et al. Use of human immunodeficiency virus-1 protease inhibitors is associated with atherogenic lipoprotein changes and endothelial dysfunction. Circulation 2001;104:257–62.
111. Tabib A, Greenland T, Mercier I, et al. Coronary lesions in young HIV-positive patients at necropsy. Lancet 1992;340:730.
112. Flynn TE, Bricker LA. Myocardial infarction in HIV-infected men receiving protease inhibitors. Ann Intern Med 1999;131:458.
113. Friis-Moller N, Sabin CA, Weber R, et al. Combination antiretroviral therapy and the risk of myocardial infarction. N Engl J Med 2003;349:1993–2003.
114. Friis-Moller N, Reiss P, El-Sadr W, et al. Exposure to PI and NNRTI and risk of myocardial infarction: results from the D:A:D study [Abstract 144]. Presented at the 13th Conference on Retrovirus and Opportunistic Infections, Denver, CO, February 5–8, 2006.

115. Expert Panel on Detection, Evaluation, and Treatment of High Blood Cholesterol in Adults. Executive Summary of the Third Report of the National Cholesterol Education Program (NCEP) Expert Panel on Detection, Evaluation, and Treatment of High Blood Cholesterol in Adults (Adult Treatment Panel III). JAMA 2001;285:2486-97.
116. Ward DJ, Curtin JM. Switch from efavirenz to nevirapine associated with resolution of efavirenz-related neuropsychiatric adverse events and improvement in lipid profiles. AIDS Patient Care STDs 2006;20:542-8.
117. Bain A, Payne K, Rahman A, Bedimo R, Busti AJ. Preliminary results from a retrospective study of the lipid lowering efficacy and safety of switching within non-nucleoside reverse transcriptase inhibitors in HIV-infected patients. 2007 ACCP Spring Practice and Research Forum Meeting. April 21-25, 2007, Memphis, TN.
118. Fisac C, Fumero E, Crespo M, et al. Metabolic benefits 24 months after replacing a protease inhibitor with abacavir, efavirenz or nevirapine. AIDS 2005;19:917-25.
119. Calza L, Manfredi R, Colangeli V, et al. Substitution of nevirapine or efavirenz for protease inhibitors versus lipid-lowering therapy for the management of dyslipidaemia. AIDS 2005;19:1051-8.
120. Negredo E, Ribalta J, Ferre R, et al. Efavirenz induces a striking and generalized increase of HDL-cholesterol in HIV infected patients. AIDS 2004;18:819-21.
121. Ruiz L, Negredo E, Domingo P, et al. Antiretroviral treatment simplification with nevirapine in protease inhibitor-experienced patients with HIV-associated lipodystrophy: 1-year prospective follow-up of a multicenter, randomized, controlled study. J Acquir Immune Defic Syndr 2001;27:229-36.
122. Martinez E, Conget I, Lozano L, et al. Reversion of metabolic abnormalities after switching from HIV-1 protease inhibitors to nevirapine. AIDS 1999;13:805-10.
123. Negredo E, Ribalta J, Paredes R, et al. Reversal of atherogenic lipoprotein profile in HIV-1 infected patients with lipodystrophy after replacing protease inhibitors by nevirapine. AIDS 2002;16:1383-9.
124. Busti AJ, Bedimo R, Margolis D, Hardin D. Improvement in insulin sensitivity and dyslipidemia in protease inhibitor-treated patients after switch to atazanavir/ritonavir (ATV/RTV): a prospective study using hyperinsulinemic, euglycemic clamp testing. Antivir Ther 2006;11:L37.
125. Mobius U, Lubach-Ruitman M, Castro-Frenzel B, et al. Switching to atazanavir improves metabolic disorders in antiretroviral-experience patients with severe hyperlipidemia. J Acquir Immune Defic Syndr 2005;39:174-80.
126. Busti AJ, Hall RG, Margolis DM. Review: atazanavir (Reyataz®) for the treatment of HIV infection. Pharmacotherapy 2004;24:1732-47.
127. Busti AJ, Tsikouris JP, Peeters MJ, et al. A prospective evaluation of the effect of atazanavir on the QTc-interval and QTc-dispersion in HIV-positive patients. HIV Med 2006;7:317-22.

128. Nguyen ST, Eaton SA, Bain A, et al. A multi-center retrospective study on the clinical efficacy and safety on lipid lowering effects after a switch to an atazanavir/ritonavir based HAART regimen [Poster RP273]. 41st ASHP Midyear Clinical Meeting and Exhibition. December 3-7, 2006. Anaheim, Orange County, CA.
129. Penzak SR, Chuck SK, Stajich GV. Safety and efficacy of HMG-CoA reductase inhibitors for treatment of hyperlipidemia in patients with HIV infection. Pharmacotherapy 2000;20:1066-71.
130. Visnegarwala F, Maldonado M, Sajia P, et al. Lipid lowering effects of statins and fibrates in the management of HIV dyslipidemias associated with antiretroviral therapy in HIV clinical practice. J Infect 2004;49:283-90.
131. Miller J, Brown D, Amin J, et al. A randomized, double-blind study of gemfibrozil for the treatment of protease inhibitor-associated hypertriglyceridemia. AIDS 2002;16:2195-200.
132. Aberg JA, Zackin RA, Brobst SW, et al. A randomized trial of the efficacy and safety of fenofibrate versus pravastatin in HIV-infected subjects with lipid abnormalities: AIDS Clinical Trials Group Study 5087. AIDS Res Hum Retroviruses 2005;21:757-67.
133. Badiou S, Merle De Boever C, Dupuy AM, et al. Fenofibrate improves the atherogenic lipid profile and enhances LDL resistance to oxidation in HIV-positive adults. Atherosclerosis 2004;172:273-9.
134. Manfredi R, Calza L, Chiodo F. Polyunsaturated ethyl esters of n-3 fatty acids in HIV-infected patients with moderate hypertriglyceridemia: comparison with dietary and lifestyle changes, and fibrate therapy. J Acquir Immune Defic Syndr 2004;36:878-80.
135. Henry K, Melroe H, Huebesch J, et al. Atorvastatin and gemfibrozil for protease-inhibitor-related lipid abnormalities. Lancet 1998;352:1031-2.
136. De Truchis P, Kirstetter M, Perier A, et al. Reduction in triglyceride level with N-3 polyunsaturated fatty acids in HIV-infected patients taking potent antiretroviral therapy: a randomized prospective study. J Acquir Immune Defic Syndr 2007;44:278-85.
137. Wohl D, Tien HC, Busby M, et al. Randomized study of the safety and efficacy of fish oil (omega-3 fatty acid) supplementation with dietary and exercise counseling for the treatment of antiretroviral therapy-associated hypertriglyceridemia. Clin Infect Dis 2005;41:1498-504.
138. Gerber J, Kitch D, Aberg J, et al. The safety and efficacy of fish oil in combination with fenofibrate in subjects on ART with hypertriglyceridemia who had an incomplete response to either agent alone: results of ACTG A58186 [Session 35 Oral Abstracts]. Presented at the 13th Conference on Retrovirus and Opportunistic Infections, Denver, CO, February 5-8, 2006.
139. Henry K, Melroe H, Huebesch J, et al. Atorvastatin and gemfibrozil for protease-inhibitor-related lipid abnormalities. Lancet 1998;352:1031-2.
140. Gerber MT, Mondy KE, Yarasheski KE, et al. Niacin in HIV-infected individuals with hyperlipidemia receiving potent antiretroviral therapy. Clin Infect Dis 2004;39:419-25.

141. Chastain L, Bain A, Edwards K, Bedimo R, Busti AJ. A retrospective study of the lipid lowering efficacy and safety of ezetimibe added to HMG-CoA reductase therapy in HIV-infected patients with hyperlipidemia. 2007 ACCP Spring Practice and Research Forum Meeting. April 21–25, 2007, Memphis, TN.
142. Coll B, Aragones G, Parra S, et al. Ezetimibe effectively decreases LDL-cholesterol in HIV-infected patients. AIDS 2006;20:1675–7.
143. Negredo E, Molto J, Puig J, et al. Ezetimibe, a promising lipid-lowering agent for the treatment of dyslipidaemia in HIV-infected patients with poor response to statins. AIDS 2006;20:2159–64.
144. Van der Lee MJ, Vogel M, Schippers E, et al. Pharmacokinetics of combined use of lopinavir/ritonavir and rosuvastatin in HIV-infected patients [Abstract 588]. Presented at the 13th Conference on Retrovirus and Opportunistic Infections, Denver, CO, February 5–8, 2006.
145. Gerber JG, Rosenkranz SL, Fichtenbaum CJ, et al. Effect of efavirenz on the pharmacokinetics of simvastatin, atorvastatin, and pravastatin. Results of AIDS Clinical Trials Group 5108 Study. J Acquir Immune Defic Syndr 2005;39:307–12.
146. Raltegravir (Isentress) [package insert]. Whitehouse Station, NJ: Merck & Co.; 2007.
147. Wenke K. Management of hyperlipidemia associated with heart transplantation. Drugs 2004;64:1053–68.
148. Bilchick KC, Henrikson CA, Skojec D, et al. Treatment of hyperlipidemia in cardiac transplant recipients. Am Heart J 2004;148:200–10.
149. Ojo AO. Cardiovascular complications after renal transplantation and their prevention. Transplantation 2006;82:603–11.
150. Perrault LP, Mahlberg F, Breugnot C, et al. Hypercholesterolemia increases coronary endothelial dysfunction, lipid content and accelerated atherosclerosis after heart transplantation. Arterioscler Thromb Vasc Biol 2000;20:728–36.
151. Esper E, Glagov S, Karp RB, et al. Role of hypercholesterolemia in accelerated transplant coronary vasculopathy: results of surgical therapy with partial ileal bypass in rabbits undergoing heterotopic heart transplantation. J Heart Lung Transplant 1997;16:420–35.
152. National Kidney Foundation. Clinical practice guidelines for managing dyslipidemias in kidney transplant patients: a report from the Managing Dyslipidemias in Chronic Kidney Disease Work Group of the National Kidney Foundation Kidney Disease Outcomes Quality Initiative. Am J Transplant 2004;4(suppl 7):13–53.
153. Azathioprine (Azasan) [package insert]. Morrisville, NC: Salix Pharmaceuticals; 2003.
154. Cyclosporine (Gengraf) [package insert]. North Chicago, IL: Abbott Laboratories; 2004.
155. Mycophenolate mofetil (cellcept) [package insert]. Nutley, NJ: Roche Pharmaceuticals; 2005.
156. Sirolimus (Rapamune) [package insert]. Philadelphia: Wyeth Pharmaceuticals; 2007.

157. Tacrolimus (Prograf) [package insert]. Deerfield, IL: Astellas Pharma US; 2006.
158. Ballantyne CM, Radovancevic B, Farmer JA, et al. Hyperlipidemia after heart transplantation: report of a 6-year experience with treatment recommendations. J Am Coll Cardiol 1992;19:1315–21.
159. White M, Ross H, Haddad H, et al. Subclinical inflammation and prothrombotic state in heart transplant recipients: impact of cyclosporin microemulsion vs. tacrolimus. Transplantation 2006;82:763–70.
160. Renlund DG, Bristow MR, Crandall BG, et al. Hypercholesterolemia after heart transplantation: amelioration by corticosteroid-free maintenance immunosuppression. J Heart Ling Transplant 1989;8:214–9.
161. Keogh A, Macdonald P, Harvison A, et al. Initial steroid-free versus steroid-based maintenance therapy and steroid withdrawal after heart transplantation: two views of the steroid questions. J Heart Lung Transplant 1992;11:421–7.
162. Vanrenterghem Y, Lebranchu Y, Hené R, Oppenheimer F, Ekberg H. Double-blind comparison of two corticosteroid regimens plus mycophenolate mofetil and cyclosporine for prevention of acute renal allograft rejection. Transplantation 2000;70:1352–9.
163. Hollander AA, Hene RJ, Hermans J, Van Es LA, van der Woude FJ. Late prednisone withdrawal in cyclosporine-treated kidney transplant patients: a randomized study. J Am Soc Nephrol 1997;8:294–301.
164. Kupin W, Venkat KK, Oh HK, Dienst S. Complete replacement of methylprednisolone by azathioprine in cyclosporine-treated primary cadaveric renal transplant recipients. Transplantation 1988;45:53–5.
165. Ingulli E, Tejani A, Markell M. The beneficial effects of steroid withdrawal on blood pressure and lipid profile in children posttransplantation in the cyclosporine ERA. Transplantation 1993;55:1029–33.
166. Hilbrands LB, Demacker PN, Hoitsma AJ, Stalenhoef AF, Koene RA. The effects of cyclosporine and prednisone on serum lipid and (apo)lipoprotein levels in renal transplant recipients. J Am Soc Nephrol 1995;5:2073–81.
167. Johnson C, Ahsan N, Gonwa T, et al. Randomized trial of tacrolimus (Prograf) in combination with azathioprine or mycophenolate mofetil versus cyclosporine (Neoral) with mycophenolate mofetil after cadaveric kidney transplantation. Transplantation 2000;69:834–41.
168. McCune TR, Thacker LR II, Peters TG, et al. Effects of tacrolimus on hyperlipidemia after successful renal transplantation: a Southeastern Organ Procurement Foundation multicenter clinical study. Transplantation 1998;65:87–92.
169. Taylor DO, Barr ML, Radovancevic B, et al. A randomized, multicenter comparison of tacrolimus and cyclosporine immunosuppressive regimens in cardiac transplantation: decreased hyperlipidemia and hypertension with tacrolimus. J Heart Lung Transplant 1999;18:336–45.
170. Groth CG, Backman L, Morales JM, et al. Sirolimus (rapamycin)-based therapy in human renal transplantation: similar efficacy and different toxicity compared with cyclosporine. Sirolimus European Renal Transplant Study Group. Transplantation 1999;67:1036–42.

171. Hoogeveen RC, Ballantyne CM, Pownall HJ, et al. Effect of sirolimus on the metabolism of ApoB100-containing lipoproteins in renal transplant patients. Transplantation 2001;72:1244–50.
172. Asberg A, Hartmann A, Fjeldsa E, Bergan S, Holdaas H. Bilateral pharmacokinetic interaction between cyclosporine A and atorvastatin in renal transplant recipients. Am J Transplant 2001;1:382–6.
173. Goldberg R, Roth D. Evaluation of fluvastatin in the treatment of hypercholesterolemia in renal transplant recipients taking cyclosporine. Transplantation 1996;62:1559–64.
174. Velosa JA, La Belle P, Ronca PD, et al. Pharmacokinetics of lovastatin in renal transplant patients on azathioprine or cyclosporine. J Am Soc Nephrol 1990;1:325.
175. Olbricht C, Wanner C, Eisenhauer T, et al. Accumulation of lovastatin, but not pravastatin, in the blood of cyclosporine-treated kidney graft patients after multiple doses. Clin Pharmacol Ther 1997;62:311–21.
176. Arnadottir M, Eriksson L-O, Thysell H, Karkas JD. Plasma concentration profiles of simvastatin 3-hydroxy-3-methyl-glutaryl-coenzyme A reductase inhibitory activity in kidney transplant recipients with and without cyclosporin. Nephron 1993;65:410–3.
177. Ichimaru N, Takahara S, Kokado Y, et al. Changes in lipid metabolism and effect of simvastatin in renal transplant recipients induced by cyclosporine or tacrolimus. Atherosclerosis 2001;158:417–23.
178. Lemahieu WPD, Hermann M, Asberg A, et al. Combined therapy with atorvastatin and calcineurin inhibitors: no interactions with tacrolimus. Am J Transplant 2005;5:2236–43.
179. Buchanan C, Smith L, Corbett J, Nelson E, Shihab F. A retrospective analysis of ezetimibe treatment in renal transplant recipients. Am J Transplant 2006;6:770–4.
180. Kohnle M, Pietruck F, Kribben A, et al. Ezetimibe for the treatment of uncontrolled hypercholesterolemia in patients with high-dose statin therapy after renal transplantation. Am J Transplant 2006;6:205–8.
181. Langone AJ, Chuang P. Ezetimibe in renal transplant patients with hyperlipidemia resistant to HMG-CoA reductase inhibitors. Transplantation 2006;81:804–7.
182. Keogh A, Day R, Critchley L, et al. The effect of food and cholestyramine on the absorption of cyclosporine in cardiac transplant recipients. Transplant Proc 1988;20:27–30.
183. Boissonnat P, Salen P, Guidollet J, et al. The long-term effects of the lipid-lowering agent fenofibrate in hyperlipidemic heart transplant recipients. Transplantation 1994;58:245–7.
184. Chan TM, Cheng IK, Tam SC. Hyperlipidemia after renal transplantation: treatment with gemfibrozil. Nephron 1994;67:317–21.
185. Bastani B, Robinson S, Heisler T, et al. Post-transplant hyperlipidemia: risk factors and response to dietary modification and gemfibrozil therapy. Clin Transplant 1995;9:340–8.
186. Prueksaritanont T, Zhao JJ, Ma B, et al. Mechanistic studies on metabolic interactions between gemfibrozil and statins. J Pharm Exp Ther 2002;301:1042–51.
187. Prueksaritanont T, Tang C, Qiu Y, et al. Effects of fibrates on metabolism of statins in human hepatocytes. Drug Metab Disp 2002;30:1280–7.

188. Fleischhauer FJ, Yan WD, Fischell TA. Fish oil improves endothelium-dependent coronary vasodilation in heart transplant recipients. J Am Coll Cardiol 1993;21:982-9.
189. Grimminger F, Grimm H, Fuhrer D, et al. Omega-3 lipid infusion in a heart allotransplant model. Shift in fatty acid and lipid mediator profiles and prolongation of transplant survival. Circulation 1996;93:365-71.
190. Uzark K, Crowley D, Callow L, et al. Hypercholesterolemia after cardiac transplantation in children. Am J Cardiol 1990;66:1385-7.
191. Henkin Y, Oberman A, Hurst DC, et al. Niacin revisited: clinical observations on an important but underutilized drug. Am J Med 1991;91:239-46.
192. Castelli WP, Anderson K, Wilson PW, Levy D. Lipids and risk of coronary heart disease. The Framingham Study. Ann Epidemiol 1992;2:23-8.
193. Grundy SM, Cleeman JI, Rifkind BM, Kuller LH. Cholesterol lowering in the elderly population. Coordinating Committee of the National Cholesterol Education Program. Arch Intern Med 1999;159:1670-8.
194. Malenka DJ, Baron JA. Cholesterol and coronary heart disease. The importance of patient-specific attributable risk. Arch Intern Med 1988;148:2247-52.
195. Tresch DD, Aronow WS. Clinical manifestations and diagnosis of coronary artery disease. Clin Geriatr Med 1996;12:89-100.
196. Forman R, Aronow WS. Management of postmyocardial infarction in the elderly patient. Clin Geriatr Med 1996;12:169-80.
197. Heart Protection Study Collaboration. MRC/BHF Heart Protection Study of cholesterol lowering with simvastatin in 20, 536 high-risk individuals: a randomized placebo-controlled trial. Lancet 2002;360:7-22.
198. Hunt D, Young P, Simes J, et al. Benefits of pravastatin on cardiovascular events and mortality in older patients with coronary heart disease are equal to or exceed those seen in younger patients: results from the LIPID trial. Ann Intern Med 2001;134:931-40.
199. Miettinen TA, Pyorala K, Olsson AG, et al. Cholesterol-lowering therapy in women and elderly patients with myocardial infarction or angina pectoris: findings from the Scandinavian Simvastatin Survival Study (4S). Circulation 1997;96:4211-8.
200. Lewis SJ, Moye LA, Sacks FM, et al. Effect of pravastatin on cardiovascular events in older patients with myocardial infarction and cholesterol levels in the average range. Results of the Cholesterol and Recurrent Events (CARE) trial. Ann Intern Med 1998;129:681-9.
201. Downs JR, Clearfield M, Weis S, et al. Primary prevention of acute coronary events with lovastatin in men and women with average cholesterol levels. Results from AFCAPS/TexCAPS. JAMA 1998;279:1615-22.
202. Shepherd J, Blauw GJ, Murphy MB, et al; PROSPER study group. Pravastatin in elderly individuals at risk of vascular disease (PROSPER): a randomized controlled trial. PROspective Study of Pravastatin in the Elderly at Risk. Lancet 2002;360:1623-30.
203. Deedwania P, Stone PH, Merz CN, et al. Effects of intensive versus moderate lipid-lowering therapy on myocardial ischemia in older patients with coronary heart disease. Results of the Study Assessing Goals in Elderly (SAGE). Circulation 2007;115:700-7.

204. Rubins HB, Robins SJ, Collins D, et al. Gemfibrozil for the secondary prevention of coronary heart disease in men with low levels of high-density lipoprotein cholesterol. Veterans Affairs High-Density Lipoprotein Cholesterol Intervention Trial Study Group. N Engl J Med 1999;341:410-8.
205. LaRosa JC, Vupputuri S. Effects of statins on risk of coronary disease: a meta-analysis of randomized controlled trials. JAMA 1999;282:2340-6.
206. Afilalo J, Duque G, Steele R, et al. Statins for secondary prevention in elderly patients: a hierarchical Bayesian meta-analysis. J Am Coll Cardiol 2008;51:46-8.
207. Foody JM, Shah R, Galusha D, et al. Statins and mortality among elderly patients hospitalized with heart failure. Circulation 2006;113:1086-92.
208. Alexander KP, Roe MT, Chen AY, et al. Evolution in cardiovascular care for elderly patients with non-ST segment elevation acute coronary syndromes: results from the CRUSADE National Quality Improvement Initiative. J Am Coll Cardiol 2005;46:1479-87.
209. Berger AK, Duval SJ, Armstrong C, et al. Contemporary diagnosis and management of hypercholesterolemia in elderly acute myocardial infarction patients: a population-based study. Am J Geriatr Cardiol 2007;16:15-23.
210. Brass LM, Alberts MJ, Sparks L. An assessment of statin safety by neurologists. Am J Cardiol 2006;97(suppl):86C-88C.
211. Bernick C, Katz R, Smith NL, et al. Statins and cognitive function in the elderly: the Cardiovascular Health Study. Neurology 2005;65:1388-94.
212. Setoguchi S, Glynn RJ, Avorn J, et al. Statins and the risk of lung, breast, and colorectal cancer in the elderly. Circulation 2007;115:27-33.
213. Dale KM, Coleman CI, Henyan NN, et al. Statins and cancer risk: a meta-analysis. JAMA 2006;295:74-80.
214. Lipka L, Sager P, Strony J, et al. Efficacy and safety of coadministration of ezetimibe and statins in elderly patients with primary hypercholesterolaemia. Drugs Aging 2004;21:1025-32.
215. Clinical Quality Improvement Network Investigators. Low incidence of assessment and modification of risk factors in acute care patients at high risk for cardiovascular events, particularly among females and the elderly. The Clinical Quality Improvement Network (CQIN) Investigators. Am J Cardiol 1995;76:570-3.
216. Olsson AG, Schwartz GG, Szarek M, et al. Effects of high-dose atorvastatin in patients > or = 65 years of age with acute coronary syndrome (from the myocardial ischemia reduction with aggressive cholesterol lowering [MIRACL] study). Am J Cardiol 2007;99:632-5.
217. Foody JM, Shah R, Galusha D, et al. Statins and mortality among elderly patients hospitalized with heart failure. Circulation 2006;113:1086-92.

8

CLINICAL PRACTICE PEARLS

Evan M. Sisson, Pharm.D., MSHA, CDE

INTRODUCTION

Providers caring for patients often discover strategies and practices they consider invaluable. The purpose of this chapter is to review some of the practice tips or clinical pearls associated with the management of dyslipidemia. The chapter is organized by bulleted clinical pearls, followed by supporting information, and is intended to supplement information provided in this and other sources.

NOVEL AND EMERGING RISK FACTORS

Aggressively treating patients with established CHD or a CHD risk equivalent is straightforward; however, many high-risk patients do not present with such obvious markers of disease. Several tests are available to measure novel and emerging risk factors for CHD that may be used individually or in combination to identify patients at high CHD risk.

MEASUREMENT OF VASCULAR DISEASE
- The greatest challenge of treating patients with dyslipidemia is selecting which patients need therapy least and providing it to those who need it most.

Ankle Brachial Index

Several tests have been proposed to identify the presence of vascular disease. The ankle

FIGURE 1. MEASUREMENT OF ANKLE BRACHIAL INDEX (ABI)

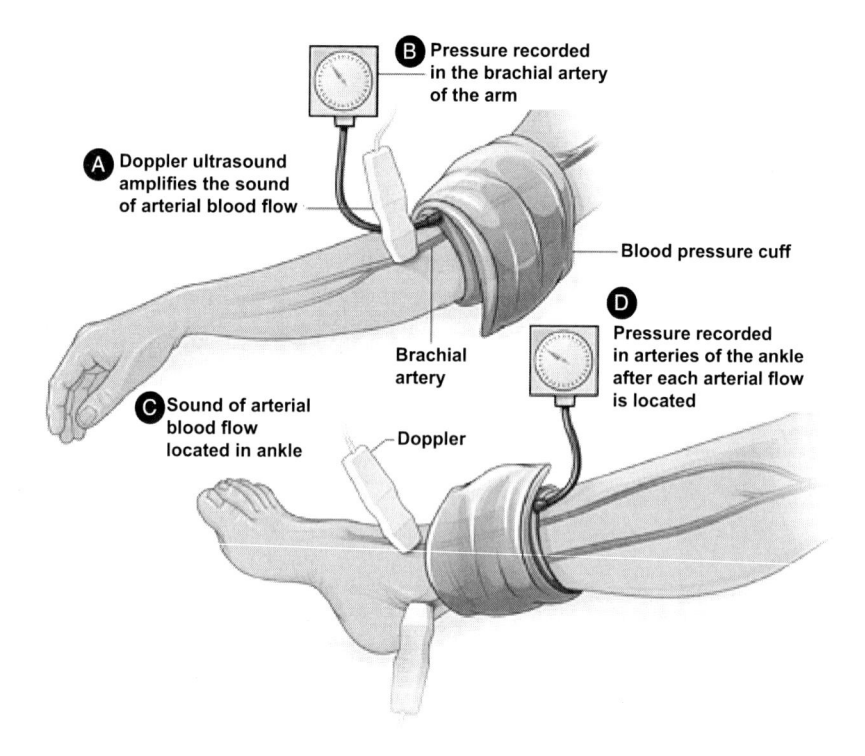

National Heart Lung and Blood Institute Diseases and Conditions Index. Available at www.nhlbi.nih.gov/health/dci/Diseases/pad/pad_diagnosis.html.

brachial index (ABI) is measured by dividing the systolic blood pressure in the ankle by that in the arm (brachial artery; see Figure 1). A ratio of less than 0.90 indicates the presence of peripheral arterial disease and an increased risk of CHD. The ABI is extremely sensitive for detecting peripheral arterial disease (up to 95% sensitivity); however, it is less sensitive for the detection of subclinical coronary disease.

Carotid B-mode Ultrasound

Carotid B-mode ultrasound evaluates the intima-media thickness in an effort to predict future risk of myocardial infarction. Increased intima-media thickness values are associated with a 5-fold increase in CHD risk. However, because of variations in methodology and limited availability, this test is currently not very useful clinically.

Exercise Tolerance Test

The exercise tolerance test is well established as a means to evaluate reversible angina. In addition to EKG changes, observations of test duration and intensity may provide insight into subclinical disease and predict future coronary events. Unfortunately, the sensitivity of the exercise tolerance test is limited and, in an asymptomatic patient, may require greater than 60% stenosis in a major coronary artery to generate an abnormal test.

Electron Beam Computed Tomography
To detect subclinical disease not observed by the exercise tolerance test, some clinicians might choose electron-beam computed tomography (EBCT) to evaluate patients. This test measures the extent of calcium deposition in coronary vessels and compares it with a database of scores correlated with atherosclerosis as determined by autopsy and angiography. A coronary calcium score above the 75th percentile for age suggests advanced atherosclerotic disease and qualifies the patient for intensive lipid-lowering therapy. As with other high-tech visualization procedures, EBCT is not widely available but could be useful in patients whose plaque burden is more extensive than might otherwise be predicted by age or other risk factors.[1]

ADVANCED LIPID TESTING METHODS AND EMERGING RISK FACTORS
Vertical Density Gradient Ultracentrifugation by Atherotech (http://www.thevaptest.com/)
- Vertical density gradient ultracentrifugation is the least expensive extended laboratory test and useful to determine LDL-C particle size.

Vertical density ultracentrifugation was developed by Jere Segrest at the University of Alabama and uses a discontinuous gradient formed by potassium bromide and a normal saline solution. Plasma samples placed in the ultracentrifugation tube spin at 65,000 rpm for 47 minutes. Because of differences in density and volume, VLDL floats to the top, whereas HDL-C remains at the bottom of the tube with LDL-C in between. The contents of the tube are subjected to an enzymatic reagent and passed through a photodensitometer to quantify the stain. The results are divided into six groupings of cholesterol concentrations corresponding to VLDL, IDL, LDL-C, Lp[a], HDL_2, and HDL_3.

Segmented Gradient Gel Electrophoresis by Berkeley HeartLab (http://www.bhlinc.com/)
- Gradient gel electrophoresis is expensive, but it is considered the gold standard methodology to determine particle size.

A second method to separate lipoproteins by size uses electrophoresis in nondenaturing polyacrylamide gels. A matrix of cross-links in the gel traps lipoproteins of different sizes as they travel from one pole to the other. This process allows sample separation of LDL-C into 7 subclasses and 10 different subclasses of HDL-C. In addition to LDL-C and HDL-C subclasses, LDL-C size is reported for the two predominant LDL-C peaks with optional reporting of apoB to indicate particle number, apoE phenotype, and Lp(a).

Nuclear Magnetic Resonance LipoProfile by LipoScience (www.liposcience.com/)
- Nuclear magnetic resonance (NMR) is more expensive than vertical ultracentrifugation and less well established than gradient gel electrophoresis.

The newest method involves proton NMR spectroscopy. Plasma samples are subjected to an NMR pulse, and the spectrum emitted by the terminal methyl group of cholesterol, cholesterol esters, triglycerides, and phospholipids is measured. The process is like measuring the sound of a bell choir. The choir produces an overall sound, but each individual bell contributes a specific tone to the chord. The intensity of the tone is related to the number of bells of a similar size. Through previous identification of signal amplitude and frequency from lipoprotein particles of known size and type, computerized algorithms can convert the overall signal response into identification and number of three different sizes for LDL-C: one for IDL, five for HDL-C, and six for VLDL. The report for LDL-C particle number includes a risk determination based on percentiles within the Framingham Offspring Population. Patients with the lowest Framingham risk category (20th percentile) have an LDL-C particle number of less than 1100 nmol/L, which corresponds to an LDL-C concentration of less than 100 mg/dL. Conversely, patients in the 80th percentile have an LDL-C particle number of 1800 nmol/L, which corresponds to an LDL-C concentration of about 180 mg/dL (calculated by dividing the particle number by 10). Although very good at measuring the size and number of cholesterol-containing particles, NMR is unable to distinguish between LDL-C and Lp(a).

Plasma Lp(a)
- Plasma Lp(a) is present in patients with CHD but may not be responsible for the disease.

Lipoprotein(a) is a complex molecule composed of LDL-C linked to a large glycoprotein (apoA). Epidemiology studies have correlated elevated concentrations of Lp(a) with increased CHD risk, but a causal link has not yet been found. In the atherosclerotic plaques, Lp(a) is thought to interact with proteoglycans in the arterial matrix and to promote inflammation after oxidative modification. Women with CHD tend to have higher concentrations of Lp(a), especially after menopause or bilateral oophorectomy. In one study, elevated Lp(a) was reduced by 30% after 4 months of treatment with 0.625 mg of conjugated estrogen. Other pharmacological agents such as niacin and neomycin have also been reported to lower Lp(a) concentrations; however, no data exist to prove that lowering Lp(a) with any intervention reduces the risk of CHD.

Homocysteine
- Although lowering homocysteine concentrations may not prevent CHD, increasing folic acid intake has few adverse effects.

Homocysteine is usually not included in local lipid panels; however, it is often offered as an optional test from extended testing services. Homocysteine is a sulfur-containing amino acid formed during the metabolism of methionine from dietary protein. In the AFCAPS/TexCAPS trial, patients with average cholesterol and low HDL-C experienced a 15% increased risk of acute coronary events with each increasing quartile of baseline homocysteine. Women in this highest quartile of elevated homocysteine at baseline in the Women's Health Study were twice as likely to experience a coronary event versus those in the lowest quartile. Homocysteine is thought to exert its atherogenic risk by several mechanisms. Once in the plasma, homocysteine is rapidly auto-oxidized to form homocystine and homocysteine thiolactone. This latter product aggregates with LDL-C particles, leading to the enhanced uptake of LDL-C particles by macrophages in the formation of foam cells. Hydrogen peroxide formed during the auto-oxidation of homocysteine causes proliferation of smooth muscle cells (which promote the activation of platelets and leukocytes as described earlier), LDL-C oxidation, and endothelial dysfunction.

Elevated concentrations of homocysteine are related to genetics, gender, age, lifestyle, drug use, and concomitant diseases. Treatment of hyperhomocysteinemia involves interventions of the modifiable risk factors. Several studies, including the Nurses Health Study and Framingham Heart Study, have shown vitamin replacement of deficient enzyme cofactors is generally effective in lowering plasma homocysteine concentrations. Unfortunately, no prospective data exist to determine the minimal effective dose of vitamin supplementation necessary to control homocysteine concentrations. The AHA recommends increased intake of folic acid (400 mcg), vitamin B_6 (2 mg), and vitamin B_{12} (6 mcg) through increased vitamin-fortified foods or vitamin supplementation for patients at high risk of CHD. These foods include vegetables, fruits, legumes, meats, fish, and fortified grains and cereals. This treatment strategy is assisted by a recent FDA regulation requiring that all "enriched" cereal grains be fortified with 1.4 mg/kg of folic acid to prevent neural tube birth defects. The population effect of this regulation on homocysteine concentrations in patients with CHD is yet to be determined; however, a recent meta-analysis reported that folic acid supplementation does not reduce the risk of CVD or all-cause mortality in patients with a previous history of vascular disease.[2]

Apolipoprotein B
- Patients at highest CHD risk may benefit from applying non-HDL goals of therapy.

Apolipoprotein B is a marker for all of the potentially atherogenic lipoproteins, such as LDL-C, IDL, VLDL, and Lp(a). Because apoB is directly measured, it is more reliable in the setting of elevated triglyceride concentrations. Several small studies suggest that measurement of particle number by apoB better predicts CHD risk than calculated LDL-C. Data from the Atorvastatin Comparative Cholesterol Efficacy and Safety Study (ACCESS) Group indicate that apoB may actually be superior to LDL-C in predicting CHD risk. Patients (3916) in the study were randomized to one of five statin therapies and monitored for 54 weeks. Statin doses were titrated upward to the maximum (atorvastatin [10–80 mg/day]; fluvastatin (20–40 mg/day, or 40 mg twice daily); lovastatin (20–40 mg/day, or 40 mg twice daily); pravastatin (10–40 mg/day); or simvastatin (10–40 mg/day) to achieve NCEP ATP II LDL targets of therapy. A surrogate for apoB is the calculation of non-HDL. In ACCESS, fewer patients achieved non-HDL targets of therapy than ATP II LDL targets. This difference implies that patients at highest risk may benefit from applying non-HDL goals, which may demand therapy that is more aggressive. The ATP III guidelines recommend using non-HDL as a goal of therapy for patients with elevated triglyceride concentrations (greater than 250 mg/dL).

HMG-COA REDUCTASE INHIBITORS (STATINS)

DRUG SELECTION

Most of the lipid-lowering effect with statins occurs at the lower end of the dosing range (see Chapter 5, Table 1 and Figure 2).

Myositis with statins is dose related and occurs most often at the maximum dose of all statins.

In 2004, the NCEP ATP III reiterated that LDL-C is the primary target of therapy. These recommendations also emphasized that, in high-risk patients, therapy doses should be optimized not only to reach LDL-C goals, but also to be consistent with a 30–40% reduction in LDL-C. Although several statins can achieve a 30% reduction in LDL-C at maximum doses, all statins are associated with dose-related myositis.[3] The ideal statin achieves the desired LDL-C at the lower end of its dosing range.

ADMINISTRATION

- It is better for patients to take statins in the morning than not at all.

Hepatic cholesterol production generally increases overnight between midnight and 3:00 AM. Statins with very short half-lives such as lovastatin, simvastatin, and fluvastatin exhibit greater lipid-lowering effects if dosed in the evening.[4] Lovastatin is best dosed with the evening meal because fat from the meal facilitates absorption. All other currently available statins may be dosed at any time of the day without regard to meals. Because the half-life of up-regulated LDL-C receptors is 24 hours, dosing the more potent statins in the morning makes little difference. If statins are the only prescription drug dosed in the evening, then adherence to the regimen will likely be low. For many patients, the convenience of taking statins in the morning with their other drugs outweighs the small benefit from bedtime dosing.

TABLE 1. NEW DEFINITIONS TO DESCRIBE MUSCLE FINDINGS IN PATIENTS TAKING STATINS

- Myopathy[a]
  - Complaints of myalgias (muscle pain or soreness), weakness, and/or cramps, *plus*
  - Elevation in serum CK > 10 times the ULN
- Rhabdomyolysis
  - CK > 10,000 IU/L, *or*
  - CK > 10 times the ULN plus an elevation in SCr or medical intervention with IV hydration therapy[b]

---

[a]A patient may describe intolerable muscle symptoms but not be found to have a CK concentration > 10 times the ULN. This patient may be considered to be experiencing myopathy for the purpose of further evaluation.
[b]The CK concentration may be < 10 times the ULN depending on the temporal relation between the event and the drawing of the laboratory sample.
CK = creatine kinase; IV = intravenous; ULN = upper limit of normal.
Adapted from McKenney JM, Davidson MH, Jacobson TA, Guyton JR; National Lipid Association Statin Safety Assessment Task Force. Final conclusions and recommendations of the National Lipid Association Statin Safety Assessment Task Force. Am J Cardiol 2006;97:89C–94C.

ADVERSE EFFECTS

- The risk of myositis is roughly the same for all statins, but it may occur less often with pravastatin and fluvastatin.

Myalgias (muscle pain or soreness) without CK elevations are common with all statins; however, they are "rarely severe, and very rarely progress to a life-threatening situation."[3] Simvastatin 80 mg is associated with the highest incidence of rhabdomyolysis and differs from lovastatin by only one methyl group. Lipophilic drugs such as lovastatin and simvastatin may penetrate muscle more readily than hydrophilic drugs (pravastatin and rosuvastatin); however, the hydrophilic agents do not entirely eliminate the risk of myositis. Although the incidence of rhabdomyolysis is lower with pravastatin and fluvastatin, this is more likely because of their weak ability to affect HMG-CoA than their lipophilicity. If a patient on simvastatin complains of muscle pain, then switching to the more hydrophilic pravastatin might solve the problem. However, if a patient on pravastatin reports the same complaint, then selection of a low-dose synthetic statin (fluvastatin) might be an option.

DRUG INTERACTIONS

- A little grapefruit juice goes a long way.

For many years, pharmacists have been aware that large quantities (more than 1 quart/day) of grapefruit juice affect CYP3A4 in the small intestine. The flavonoids in grapefruit juice, not other fruit juices, can inhibit the clearance of certain statins, leading to increased serum concentrations and an increased risk of serious toxicity (e.g., myopathy). Concomitant administration of as little as 8 oz of grapefruit juice can increase serum simvastatin concentrations 5- to 15-fold; however, 90% of this inhibition dissipates within 24 hours.[5,6] Lovastatin, simvastatin, and atorvastatin are also inhibitors for P-glycoprotein and substrates for this efflux protein, which is located in the intestines, placenta, kidneys, brain, and liver. Grapefruit juice, St. John's wort, and other food supplements may interact with statins through both CYP3A4 and P-glycoprotein. Initiation of statin therapy in patients taking inhibitors of CYP3A4 and P-glycoprotein should be done at lower doses or with fluvastatin, pravastatin, or rosuvastatin, which appear to be less affected.

MONITORING

- Measuring CK is not useful in asymptomatic patients.[3]

Statin-induced myositis is a common finding but difficult to distinguish from normal aches and pains. Patients at higher risk of myositis are typically small-framed and older, with impaired renal function. Painful joints are typically associated with arthritic causes, whereas muscle complaints located in the belly of larger muscles are likely statin induced. True myopathy, defined as muscle pain (especially when bilateral proximal), weakness, or cramps with an serum CK greater than 10 times the ULN, is rare, and occurs in 0.08% of patients (see Table 1).[3] Routine measurement of CK concentrations in asymptomatic patients is typically not useful because of the high sensitivity and low specificity of the test. Randomly sampling patients who have an especially large muscle mass, recently exercised, or experienced other muscle trauma reveals elevated CK concentrations. In these patients, the provider must decide whether to continue statin therapy based solely on one laboratory test and no clinical symptoms. The converse of this situation also exemplifies the inadequacy of routine CK screening. If a patient who recently started a statin complains of severe muscle pain on exertion, then the clinical signs obviate the need for further testing. Even if the true cause of the muscle pain is unrelated to the statin, the patient still expects the therapy to change.

- Change in liver function tests on a stable statin dose in the absence of other causes (such as decompensated heart failure) is a red flag.

The symptoms of hepatitis induced by statins resemble the flu and include fatigue, sluggishness, anorexia, and weight loss. Aminotransferase concentrations are usually only moderately elevated (2–3 times the ULN); otherwise, there is little guidance offered from statin product labeling. An LDL-C concentration drawn at the same time as the liver function test that is much lower than expected is also a sign of toxicity. If a patient who has been well controlled on a particular statin dose suddenly exhibits an increase in enzyme concentrations 3 times above the patient's normal value, the provider should respond first by rechecking the values.[3] If the second set of values confirms the first, the statin should be discontinued for 1 week and the enzyme concentrations rechecked. If the values resolve, then an alternative therapy may be selected. This conservative approach to statin monitoring is especially recommended for patients on combination therapy or on concurrent therapy with drugs dependent on CYP elimination.

- Lowering cholesterol with statins is not associated with a mortality J-curve.

Early epidemiologic studies suggested that lowering cholesterol values beyond certain concentrations were associated with increased non-CHD mortality including accidents, violent death, cancer, and chronic obstructive pulmonary disease. Other population studies have demonstrated that a large proportion of patients who die from non-CHD events and low cholesterol concentrations are usually elderly with declining health and a low socioeconomic status. A reanalysis of the original cholesterol-lowering trials concluded that cholesterol reduction is beneficial in reducing CHD mortality and total mortality. Furthermore, the observed non-CHD deaths clustered around specific interventions such as fibrates (clofibrate and gemfibrozil), estrogen, and dextrothyroxine. If the J-curve for cholesterol truly existed, then increases in total mortality would be expected with more aggressive lipid lowering. The WHO and Helsinki Heart Study achieved a 9–10% reduction in total cholesterol, whereas the 4S, WOSCOPS, and LIPID trials achieved an 18–29% reduction. Despite much better cholesterol reduction, a decreased total mortality rather than an increased mortality has routinely been reported in the new larger trials.

- The LDL-C of children may approximate ideal normal values for adults.

The optimal LDL-C value is much lower than previously acknowledged. In primitive hunter-gatherer tribes, plasma LDL-C concentrations are reported to be 50–80 mg/dL with little or no CHD. Human newborn babies have average LDL-C concentrations of 57 mg/dL (range, 26–123 mg/dL) and a very low risk of CHD. Patients with hypobetalipoproteinemia have LDL-C values around 10 mg/dL and a normal life expectancy. These data support findings from recent statin trials, which show

no significant difference in noncardiac death between patients on therapy at various LDL-C concentrations.
- The optimal time to recheck laboratory values after starting a statin is 6 weeks to avoid subjecting patients to a second venipuncture.

To monitor for hepatotoxicity, liver function tests should be checked at baseline or after a dose increase and then again 12 weeks later and periodically thereafter.[3] Because week 6 on statin therapy likely represents the nadir of LDL-C before the patient's body compensates, providers may consider testing for effect and toxicity simultaneously. If the patient is not at goal by the 6-week follow-up, then further dose adjustments should be made, and the concurrent liver function tests may be used to verify continuation at baseline. This cycle is repeated after each dose adjustment because statin-induced hepatotoxicity is dose related. Once a patient has stabilized, periodic monitoring may be instituted; however, frequent, routine measurement of liver function tests is not recommended. Although the risk of hepatotoxicity decreases with time, the addition of interacting drugs, disease progression, and patient compliance should be monitored regularly.

- Biannual visits are best for patients on lipid-lowering therapy.

Even the most responsible patients experience deterioration in their lipid profile. If placed on a once-yearly visit schedule, the provider always has to make up ground, necessitating that the longest distance between appointments is 6 months.

FIBRIC ACID DERIVATIVES (FIBRATES)

FIBRATE SELECTION

- Combination of statins with fibrates is often necessary in patients with mixed dyslipidemia.

The complementary action of statins and fibrates makes them ideally suited to address mixed dyslipidemia; however, concerns about increased risk of myopathy with the combination limit its use. The mechanism of increased myopathy was originally thought to be a fibrate class effect related to the CYP system; however, more recently, the interaction with statins appears to be isolated to the ability of gemfibrozil to inhibit statin glucuronidation. Consequently, the maximum dose of simvastatin, rosuvastatin, and lovastatin when used with gemfibrozil should not exceed 10, 10, and 20 mg, respectively, to avoid statin accumulation and increased risk of myopathy. Clarification of this mechanism has renewed interest in using statins in combination with fenofibrate, which does not affect statin glucuronidation. The new micronized formulation of fenofibrate also allows dosing without regard to meals (starting dose is 48–145 mg/day).

INTESTINAL-ACTING AGENTS

Because most statin-related adverse effects are dose-dependent, combination therapy with agents having a different mechanism (especially cholesterol absorption blockers) is often useful to allow patients to achieve LDL-C goals. In combination with statins, colesevelam (3.8 g/day) provides an additional 10–16% reduction in plasma LDL-C from the baseline value. Gastrointestinal adverse effects with colesevelam are less than with colestipol or cholestyramine, but they are still common (constipation in 10% of patients), leading to discontinuation rates in up to 50% of patients.

BILE ACID SEQUESTRANTS

Drug Selection

- Consider BASs if safety is a concern.

The great advantage of BASs is their safety profile. The absence of systemic absorption makes BASs ideal for patients intolerant of statins or those in whom statin therapy is contraindicated. Bile acid sequestrants are the one lipid-lowering drug therapy safe for pregnant women, and they are often used in children. Cholestyramine therapy can cause hyperchloremic acidosis in children because chloride ions are released in exchange for bile acids. Bile acid sequestrants can also decrease the absorption of vitamin D and other fat-soluble vitamins, which may be significant in children. This latter concern is alleviated by vitamin supplementation, though many providers avoid this issue and choose a statin with FDA-approved pediatric labeling (lovastatin, simvastatin, pravastatin, and atorvastatin).[7]

Administration

- Many of the unpleasant issues associated with resin therapy can be overcome.

The main problem with BASs is their poor mouth feel and significant GI adverse effects. Patients who take BASs often refer to them as "sand." Drinking sand suspended in liquid occasionally happens when enjoying a day at the beach, but it does not engender great compliance when part of a therapeutic regimen. To overcome the issues related to the dosage form, patients suggest several strategies. Adding the resin to pulpy orange juice or applesauce tends to mask the grittiness, and adding the drug to warm prune juice decreases constipation and the gritty impact of resin in the stool. Some patients suggest adding resin to health shakes in place of wheat germ. Aside from the poor mouth feel, the GI complaints include abdominal pain, belching, bloating, constipation, gas, heartburn, and nausea. These adverse effects can be alleviated by having patients increase soluble fiber intake through either dietary changes or therapeutic supplements such as a glass of prune juice or three teaspoonfuls of psyllium.

Drug Interactions

- Morning dosing before the first bowel movement may provide maximum BAS effect.

Some investigators suggest that the best time to administer a resin dose is in the morning when maximal bile acid synthesis occurs. Often, patients prefer to take the entire dose in the morning for convenience and to avoid the gritty mouth feel later in the day. Compared with the powder BASs, colesevelam is better tolerated, especially when given in divided doses. Whatever the patient decides, clear instructions should be provided to separate resin administration from other susceptible drugs because of potential drug adsorption by the resin. Bile acid sequestrants should be administered at least 1 hour before or 4 hours after polar compounds such as warfarin, theophylline, digoxin, levothyroxine, and statins. Another advantage of colesevelam over other BASs is that it does not affect the bioavailability of digoxin, warfarin, or statins.

Monitoring
- Adding a resin to patients' therapy with mixed dyslipidemia may increase triglycerides.

Bile acid sequestrants bind bile acids by exchanging an anion for bile acids in the GI tract. This interruption of the enterohepatic circulation causes the liver to convert more of the hepatic cholesterol pool into bile acids. In addition to drawing in more circulating particles from the plasma, interruption of the enterohepatic circulation with resin therapy also stimulates cholesterol synthesis, resulting in increased VLDL production. This rise in triglycerides tends to be about 7% but is especially pronounced in patients with mixed dyslipidemia on resin monotherapy.

EZETIMIBE
- Addition of ezetimibe to a statin provides a better LDL-C reduction than increasing the statin dose (see Table 2).

Doubling the dose of any statin provides only an additional 6% reduction in LDL-C. Adding ezetimibe to statin therapy generally adds an 18% additional reduction in LDL-C. Coadministration of ezetimibe (10 mg/day) with low-dose statins produces reductions in LDL-C almost identical to maximal-dose statin monotherapy (46% vs. 45% with simvastatin 80 mg/day and 53% vs. 54% with atorvastatin 80 mg/day).

Recheck liver function tests after adding ezetimibe to a statin.

Adding ezetimibe to a statin may result in mild, transient, reversible elevations in transaminase concentrations. Although this rise is typically not clinically significant, providers are encouraged to follow monitoring recommendations found in the product labeling.

STEROLS AND STANOLS
- Use of plant sterols and stanols in clinical practice is very difficult and may not yield the same cholesterol reductions found in controlled trials.

TABLE 2. COMPARISON OF AVERAGE LDL-C REDUCTION WITH STATIN THERAPY[a]

Daily Dose	Rosuvastatin	Atorvastatin	Simvastatin	Lovastatin[b]	Pravastatin	Fluvastatin[b]
5 mg	45%		26%			
10 mg	52%	39%	30%	21% (23.8%)	22%	
20 mg	55%	43%	38%	27% (29.6%)	32%	22%
40 mg	63%	50%	41%	31% (35.8%)	34%	25%
60 mg				N/A (40.8%)[c]		
80 mg		60%	47%	37%		N/A (35%)[c]

[a]Average LDL-C reduction in patients with primary hypercholesterolemia on monotherapy based on FDA product labeling for each compound.
[b]Extended-release formulation shown in parenthesis.
[c]N/A = not applicable to immediate-release dosage form.

Esterified plant sterols and stanols easily dissolve in oils and margarines and are recommended by the ATP III as an adjunct to maximal diet therapy to lower LDL-C. After ingestion, esterified stanols undergo hydrolysis in the intestine, releasing unesterified stanols that compete with dietary cholesterol for micelle incorporation, thus limiting the amount of cholesterol presented to enterocytes. Clinical trials with plant stanol esters indicate that 1–3 g/day reduces total cholesterol and LDL-C by 8% and 14%, respectively, and does not interfere with the absorption of fat-soluble vitamins. Outside clinical trials, achievement of these cholesterol reductions may be difficult, especially in patients who do not regularly use margarine or eat salads. Some clinicians express concern about the caloric impact and potentially contradictory message that overweight patients may receive by adding these products to their diet. Finally, there is also concern about the long-term safety of increased plant stanols in the diet of patients at low CHD risk.

NIACIN

DRUG SELECTION

- Niacin is ideal (almost) for patients with low HDL-C.

The most effective agent currently available to raise HDL-C (up to 35%) is niacin. Niacin raises HDL-C by two mechanisms. First, niacin blocks a hepatic receptor that mediates holoparticle uptake and catabolism of HDL-C, which reduces its clearance from the serum. Second, niacin inhibits hormone-sensitive lipase in adipose tissue, decreasing mobilization and delivery of fatty acids to the liver. The net reduction in serum triglyceride concentrations (up to 40%) makes HDL less triglyceride enriched, giving it less affinity for hepatic lipase, which also catabolizes HDL. Niacin also causes a shift from small, dense LDL-C to large, buoyant particles and lowers Lp(a) concentrations by about 30%. The effect on fatty acids is the likely mechanism of niacin-induced insulin resistance observed at doses above 1000 mg/day.[8,9]

- Sustained-release niacin should be avoided because of increased hepatotoxicity.

The sustained-release dosage form may lessen the vasodilation adverse effects of niacin, but it has inferior cholesterol-lowering effects and causes increased incidence of hepatotoxicity.[8,9] Hepatotoxicity with niacin is related to the amount of niacin metabolized through the high-affinity, low-capacity pathway. This pathway produces nicotinamide, which is then metabolized to a variety of pyrimidine breakdown products. Immediate-release niacin quickly saturates the nicotinamide pathway, causing a shift to metabolism by a low-affinity, high-capacity conjugation pathway. Slow release of niacin (longer than 12 hours) disallows saturation of the nicotinamide pathway, thus causing accumulation of hepatotoxic metabolites. Providers who preferentially decide to use sustained-release niacin should limit doses to less than 2000 mg once daily to minimize the amount of niacin metabolized through hepatotoxic pathways; restrict product selection to one reliable brand; and diligently monitor liver function tests. In patients receiving concomitant therapy with statins, sustained-release niacin doses should be limited to less than 1500 mg/day.

- Extended-release niacin is better tolerated than immediate-release niacin and safer than sustained-release niacin.

Extended-release niacin, or Niaspan, offers lipoprotein effects similar to immediate-release niacin with fewer flushing reactions. Patients are advised to take extended-release niacin at night with an aspirin to minimize the potential for flushing. In addition, by taking the dose at bedtime, patients may not even notice the flushing reaction during their sleep if it occurs.[8,9]

Adverse Effects
- Patients who never report a flush on niacin are probably not taking it.

Patients taking niacin, especially the immediate-release dosage form, commonly report symptoms of flushing, tingling, itching, rash, and headaches, which are thought to result from prostaglandin-mediated vasodilation. Several strategies have been proposed to reduce these effects, but the general rule is to titrate doses very slowly, allowing patients to become acclimated to the drug. Patients should be allowed to remain at lower, tolerable doses until they are ready to resume the titration schedule. Despite the weekly titration schedule, some patients still experience mild flushing effects, especially with the first higher dose. Many patients find that a meal or low-fat snack together with an aspirin 325 mg 30 minutes before the niacin dose (allow 60 minutes before niacin for enteric-coated aspirin) alleviates these symptoms. Ibuprofen 200 mg or another nonsteroidal anti-inflammatory drug (NSAID) may also be used, but aspirin 81 mg may not be enough to inhibit prostaglandin-mediated vasodilation.[8] Patients should also avoid situations that promote vasodilation just before niacin dosing, such as hot showers, spicy foods, and hot liquid intake.
- Patients who report a pruritic rash while taking niacin probably need to discontinue the drug.

This situation can occur at any time and does not appear to be dose related.

Monitoring
- Patients with severe gout or history of kidney stones are more prone to an exacerbation of those conditions with high doses of niacin.

Niacin increases uric acid concentrations and can precipitate gout flares. Formation of uric acid crystals in the kidney serves as a nidus for calcium carbonate stones.
- Patients with reflux, but no history of peptic ulcer disease, are eligible for niacin.

Niacin is contraindicated in patients with an active peptic ulcer or a significant history of peptic ulcer disease because the prostaglandin-mediated vasodilation may precipitate or worsen a bleeding event. Gastroesophageal reflux disease is not affected by the niacin-induced vasodilation.
- Niacin may worsen glycemic control, but patients with diabetes may still benefit.

Because of its effects on HDL-C, LDL-C, and triglycerides, niacin is an ideal agent for patients with the mixed dyslipidemia associated with diabetes. Unfortunately, niacin also induces insulin resistance by increasing mobilization of free fatty acids from adipose tissue. The effect of niacin on free fatty acid flux is dose-dependent and does not significantly affect blood glucose concentrations at doses less than 1000 mg/day. Patients with poorly controlled diabetes or those on maximal doses of oral hypoglycemic agents are not good candidates for niacin therapy. Patients on insulin therapy may require a dose adjustment after adding niacin but not a change in hypoglycemic therapy because insulin lacks a ceiling dose.

NONPRESCRIPTION PRODUCTS

Unfortunately, many patients in the United States tend to seek out nonprescription products rather than earnestly incorporating positive behaviors into their lives. The two main concerns with lipid-lowering dietary supplements are lack of standardized product preparation and deficit of large well-controlled studies. Pharmacists should be diligently aware of the emerging dietary supplement data and encourage patients to challenge unsubstantiated claims before spending money on agents that may do nothing or even cause harm.

FISH OIL

- Nonprescription fish oil capsules may not eliminate contamination concerns.

Diets high in fish or that include fish oil supplements have been recommended for several years by the AHA because of evidence of reduced risk of death, nonfatal coronary events, and stroke after myocardial infarction. Enthusiasm for this recommendation was dampened somewhat by concerns about increased concentrations of mercury, polychlorinated biphenyls, dioxins, and other environmental contaminants found in larger, predatory fish. Most capsules of fish oil may avoid mercury contamination because of the manufacturing process; however, lack of manufacturing regulation raises questions about true differences between sampled batches. A new FDA-approved fish oil product came onto the market in 2005. This product requires a prescription and is indicated for patients with triglyceride concentrations above 500 mg/dL. Compared with OTC products, the prescription product boasts reliable purity and increased concentrations of EPA and DHA. Over-the-counter products also contain different amounts of EPA and DHA as well as other ingredients, making product selection difficult for patients. Each capsule of the prescription fish oil contains about 465 mg of EPA and 375 mg of DHA. Omega-3 doses of only 1 g/day (dose = EPA+DHA) reduce cardiovascular events and overall mortality; however, higher doses of up to 4 g of the prescription product result in a 44.9% triglyceride reduction from baseline.[10]

- Fish oil before meals may decrease the fishy taste.

To minimize GI complaints, patients should be advised to take fish oil immediately before meals. Taking fish oil capsules or liquid after meals or on an empty stomach creates a pool of oil near the esophagus and increases the likelihood of reflux. In addition, storing fish oil capsules in the refrigerator helps maintain the integrity of the capsule and limits the fishy taste.

- Flaxseed oil tastes better than fish oil but has little effect on triglycerides.

Flaxseed is a rich source of the omega-3 fatty acid α-linolenic acid but also contains linoleic acid, an omega-6 fatty acid. Flaxseed oil contains more omega-3 fatty acid by weight than fish oil, but the plant-based fatty acids have little effect on triglycerides. Nonetheless, flaxseed oil lowers total cholesterol and decreases thrombin-stimulated platelet aggregation. One or two tablespoons of flaxseed oil daily are recommended and can be used in salad dressings or as flavoring. Note that heat and light easily degrade flaxseed oil, such that it should not be used in cooking. Ground flaxseed is the most studied dosage form and may have the greatest impact on cholesterol because of the fiber component.

ALCOHOL

- The value of alcohol to raise HDL-C remains questionable.

In observational studies, a consistent reduction in CHD has been found with moderate alcohol consumption. The controversy lies in the appearance of a J-curve between quantity of alcohol consumed and total mortality. The lowest mortality occurs at one or two drinks (12-oz bottle of beer, 4-oz glass of wine, and 0.5-oz shot of 80-proof spirits) per day. Those who never drink or drink alcohol on occasion are at higher risk or are at least less well protected. Patients who consume three or more drinks per day have increased total mortality. This progressive increase in mortality is due to diseases common in heavy drinkers such as stroke, alcoholic cardiomyopathy, cirrhosis, pancreatitis, and several kinds of cancer. The proposed protective effect of alcohol consumption comes from its ability to increase HDL concentrations, which may translate into a risk reduction of about 30–50%. The preponderance of epidemiologic data is impressive; however, the data are limited by the quality of secondary data sources and their inability to demonstrate cause and effect. Unfortunately, randomized controlled clinical trials of alcohol consumption will probably never be performed to establish a direct link to reduction in CHD risk. Based on the quality and mix of available results, neither widespread endorsement nor prohibition of alcohol is appropriate for the general population. Rather, health professionals should make individual recommendations based on the total risks and benefits for a particular patient.

VITAMINS AND SUPPLEMENTS

- Vitamin supplements may not decrease cardiovascular risk and may cause other health concerns.

Dietary antioxidants such as vitamin C (ascorbic acid), vitamin E (alpha-tocopherol), and beta-carotene (provitamin A) have received much attention in recent years. Although it is true that diets rich in fruits, vegetables, and whole grains that contain antioxidants may lower CHD risk, the benefits from antioxidant supplementation remain unclear. The correlation is confounded by the fact that foods high in vitamins C or E or beta-carotene tend also to be lower in saturated fat and cholesterol and high in fiber, all of which are known to decrease CHD risk. Vitamin E, long studied for its antioxidant properties, reduces the effectiveness of statins and niacin when given concurrently. The final evidence against vitamin E use came from the HPS and Women's Health Study. In both studies, vitamin E provided no mortality or cardiovascular event benefit over placebo, and in HPS, there was a trend toward increased events with vitamin E. Therefore, patients should not take supplemental vitamin E as a means to prevent CHD.

- Odorless garlic has little effect on cholesterol lowering.

The most commonly known herbal supplement used to lower cholesterol is garlic. When garlic is cut or crushed, the enzyme allinase is released, converting alliin into allicin. Allicin is responsible for the familiar odor and is thought to be the active pharmacological compound to inhibit cholesterol synthesis. In a meta-analysis of 16 randomized trials with 952 patients, a 12% reduction in serum cholesterol and a 13% reduction in triglycerides with garlic therapy compared with placebo were found. Unfortunately, the trials included in the meta-analysis had many methodological problems such that a clear conclusion could not be drawn without further study. Garlic preparations that purport absence of odor or taste appear to offer little benefit. Therefore, for maximum cholesterol-lowering effect, patients need to consume 4 g of fresh garlic per day (1 clove of fresh garlic or 10 mg of alliin with 4 mg of potential allicin). Because this quantity is much higher than in usual diets, patients also need to be advised of potential adverse effects such as diarrhea, flatulence, gastric irritation, heartburn, flushed face, rapid pulse, headache, and insomnia. Furthermore, because of the interference with thromboxane synthesis and platelet function, patients receiving warfarin, salicylates, or other antiplatelet agents should be warned of possible increased risk of bleeding.

- Red yeast rice should be avoided because of uncertain lovastatin concentrations.

Red yeast has been used in China for many years as a spice in traditional Chinese recipes and as a medicine. Red yeast rice contains a variety of compounds known as monacolins that inhibit HMG-CoA reductase and include monacolin K, also known as mevinolin or lovastatin. Because of its status as a food supplement, the concentration of lovastatin in commercial red yeast rice products is variable and ranges from 0.15 to 3.37 mg per capsule. The typical daily dosage is two 600-mg oral capsules twice daily with food. In one 8-week trial, total cholesterol decreased by 26%, LDL by 32.8%, and triglycerides by 19.9% from baseline. The adverse effect profile is similar to other pharmaceutical HMG-CoA reductase inhibitors and includes heartburn (1.8%), abdominal distention (0.9%), and dizziness (0.3%). Because there is no apparent intrinsic benefit versus more potent, FDA-approved HMG-CoA reductase inhibitors, red yeast rice should not be recommended. Pharmacists should also caution patients not to start red yeast rice on their own in addition to prescribed statin therapy because of the possible increased incidence of statin-induced adverse effects.

- Coenzyme Q10 prophylaxis cannot be recommended.

Anecdotal reports of lessened or improved muscle pain with coenzyme Q10 supplementation continue to spark interest in its use. One theory regarding the mechanism of statin-induced myalgias proposes that statins reduce muscle concentrations of ubiquinone. This theory has not been proved, and systematic attempts to reduce muscle symptoms with coenzyme Q10 have shown equivocal results.[3]

REFERENCES

1. Shah AM, Feldman AH, George DL, Edmundowicz D. Role of electron beam computed tomography in detecting and assessing coronary artery disease. Hosp Physician 2007;43:11–8.
2. Bazzano LA, Reynolds K, Holder KN, He J. Effect of folic acid supplementation on risk of cardiovascular diseases: a meta-analysis of randomized controlled trials. JAMA 2006;296:2720–6.
3. McKenney JM, Davidson MH, Jacobson TA, Guyton JR; National Lipid Association Statin Safety Assessment Task Force. Final conclusions and recommendations of the National Lipid Association Statin Safety Assessment Task Force. Am J Cardiol 2006;97:89C–94C.
4. Plakogiannis R, Cohen H. Optimal low-density lipoprotein cholesterol lowering—morning versus evening statin administration. Ann Pharmacother 2007;41:106–10.
5. Lilja JJ, Kivisto KT, Neuvonin PJ. Duration of effect of grapefruit juice on the pharmacokinetics of the CYP3A4 substrate simvastatin. Clin Pharmacol Ther 2000;68:384–90.
6. Holtzman CW, Wiggins BS, Spinler SA. Role of P-glycoprotein in statin drug interactions. Pharmacotherapy 2006;26:1601–7.
7. McCrindle BW, Urbina EM, Dennison BA, et al. Drug therapy of high-risk lipid abnormalities in children and adolescents: a scientific statement from the American Heart Association Atherosclerosis, Hypertension, and Obesity in Youth Committee, Council of Cardiovascular Disease in the Young, with the Council on Cardiovascular Nursing. Circulation 2007;115:1948–67.
8. Guyton JR, Bays HE. Safety considerations with niacin therapy. Am J Cardiol 2007;99(suppl):22C–31C.
9. McKenney JM. Niacin for dyslipidemia: considerations in product selection. Am J Health Syst Pharm 2003;60:995–1005.
10. Bays HE. Safety considerations with omega-3 fatty acid therapy. Am J Cardiol 2007;99(suppl):35C–43C.

9

SYSTEMATIC MANAGEMENT OF LIPID DISORDERS IN PHARMACEUTICAL CARE

Kenneth A. Kellick, Pharm.D., CLS, and Ralph La Forge, MSc, CLS

To understand the role of the pharmacist in lipid management, we need to begin 20 years ago when the definition of pharmaceutical care originated coincident with a new focus on lipid management. The classic Hepler and Strand definition, "Pharmaceutical care is the responsible provision of drug therapy for the purpose of achieving definite outcomes that improve a patient's quality of life,"[1] describes the framework on which many contemporary lipid practices are based.

How does a pharmacist begin to identify a need for and establish a lipid clinic practice? Because lipid management is innately interwoven with the concept of cardiovascular risk management and event prevention, establishment of the necessary disease base is essential. It may be helpful at the outset to define a process whereby lipid disorders are managed more systematically (i.e., a lipid clinic service: *A coordinated and systematic process whereby patients who have a lipid or lipoprotein disorder are identified, risk-triaged and expediently managed to acceptable lipid/lipoprotein and behavioral goals by a qualified and dedicated staff*). In contrast to usual care, such programs use defined treatment pathways grounded in currently published consensus diagnostic and treatment guidelines for lipid and lipoprotein disorders (e.g., NCEP ATP III, ADA 2007). Following the recommendations of the ATP III report in 2001,[2] note that, for primary prevention, the risk factors are as follows:

Nonmodifiable	Modifiable
Age[a]	Hypertension[a]
Sex[a]	Cigarette smoking[a]
Family history of premature CHD	Diabetes
	Obesity
	Physical inactivity
	HDL-C concentration
	Atherogenic diet
	Thrombogenic state

[a]Notable in the NHLBI risk calculator for high-risk noncardiac patients.[3]

A lipid clinic service can generally be categorized into one of two levels: a general lipid clinic service or a specialty lipid clinic service. The latter predominantly focuses on more difficult or complex cases. A general lipid service is perhaps the most prevalent form of systematic lipid management and focuses on the straightforward identification and management of dyslipidemia patient cases (e.g., polygenic dyslipidemia) as well as some moderately complex cases (e.g., generalized hypertriglyceridemia, diabetic dyslipidemia). Additional time and staff expense are required for a general lipid clinic service (Level II) that addresses more difficult or complex cases in which a more definitive diagnosis, more frequent follow-up, and more complex therapy are often required. Complex lipid disorders require a higher level of diagnostic and therapeutic skill including an understanding of the application of more advanced lipoprotein/apoprotein tests. Table 1 lists examples of more complex dyslipidemia/dyslipoproteinemias. Note that the categories listed in Table 1 are somewhat of an oversimplification because most of these have multiple phenotypes and genotypes.[4] Of course, a lipid clinic can offer both levels of lipid clinic services. To enhance professional development in either lipid disorder management setting, the pharmacist should consider board certification in clinical lipidology. The National Lipid Association (www.lipid.org) offers several clinical lipidology-credentialing programs, including the American Board for the Certification of Lipid Specialists for physicians and the Accreditation Council for Clinical Lipidology for qualified and experienced nonphysicians including pharmacists. Preparation for either board certification can also be obtained through the National Lipid Association's Self-Assessment Program and Masters in Lipidology

TABLE 1. EXAMPLES OF RELATIVELY COMPLEX DYSLIPIDEMIAS

1. Diabetic dyslipidemia
2. Familial hypertriglyceridemia
3. Familial combined hyperlipidemia
4. Chylomicronemia
5. Familial hypoalphalipoproteinemia
6. Heterozygous and homozygous hypercholesterolemia
7. Familial dysbetalipoproteinemia
8. Therapeutically resistant dyslipidemias

board review course. For more information on the Accreditation Council for Clinical Lipidology, refer to their Web site at www.lipidspecialist.org.[5]

APPROPRIATE ENTRY CRITERIA AND SUFFICIENT PATIENT REFERRAL FOR SPECIALIZED LIPID CLINIC SERVICES

When organizing a lipid clinic service at any level, it is paramount that consideration be given to the formal establishment of written referral criteria. Many lipid programs have no formalized or written referral criteria. Local provider referral sources should have some idea of what dyslipidemias/dyslipoproteinemias best suit the pharmacist's program's clinical management skills and health care setting.

The primary source of patients with treatable lipid disorders is the pharmacist's own medical or pharmacy group's higher-risk and/or more complex dyslipidemic patients. Patients who require relatively simple and straightforward monotherapy and moderate dietary changes generally do not require a dedicated lipid clinic service, except for perhaps a semiannual or annual follow-up to ensure compliance with lifestyle changes and/or drug therapy. Such patients represent most patients in pharmacy care dyslipidemia programs. Those at higher risk (e.g., secondary prevention with two or more CVD risk factors, that is, more than 20–30+% 10-year CHD risk) and/or those with more complex lipid disorders (Table 1) who require additional laboratory assays or multiple drug therapy are more appropriate for more specialized lipid clinic services requiring more frequent follow-up (e.g., 6- to 8-week return visit incidence). Specialized services often require more frequent laboratory monitoring, pharmacotherapy changes, and lifestyle therapy adjustments. Many of these will require a higher level of diagnostic proficiency and working knowledge of advanced lipid and lipoprotein assays (e.g., LDC-C particle concentration quantification, VLDL quantification, Lp(a), apolipoprotein testing). Patients who require more complex therapy (e.g., two or more liver metabolized drugs) are also good candidates for lipid clinic services. Special populations (e.g., PCOS, HIV) or pediatric dyslipidemia specialization and/or those who have historically been resistant or unresponsive to therapy are also candidates for a lipid clinic service. Regardless of what criteria a provider chooses to refer a patient, all new patients should receive a thorough evaluation to confirm the type and origin of the dyslipidemia/dyslipoproteinemia.

PATIENT EVALUATION

It is essential that the pharmacist be able to document the presence and nature of the above in an easily retrievable database, either electronic or paper. Because modifiable risk factors change in their parameters over time, the database must easily reflect these changes. Other data points to consider are weight, fasting blood glucose, BMI, and WC to assess metabolic syndrome. Waist circumference measures are best evaluated with professional standardized tape measures that consistently place a standard amount of pressure on each WC measurement. Gulick tape measures can be purchased from many sports medicine supply companies online.

The physical space to obtain the above information should be relatively quiet and free of interruptions and distractions, like all other counseling areas. Begin by noting vital demographics, age, sex, and menopausal status, if of the female sex. The pharmacist should then inquire about relevant instances of family history of premature CHD but pay close attention to those stories of history of premature CHD (definite myocardial infarction or sudden death before age 55 in father or other first-degree male relative or before age 65 in mother or other female first-degree relative). Family history modifies the treatment plan in terms of urgency (e.g., a patient with a strong family history of a first-degree relative with an event at a young age would require a more aggressive approach). Continue developing the baseline information by ascertaining the existence or absence of CVD. Antianginal drugs often can begin the discussion. It is most critical to document the time course of CVD. Distant events may impart a less aggressive treatment plan. The presence of other forms of cardiovascular or vascular disease (CHD and CHD equivalents) should be documented (e.g., history of stroke, transient ischemic attacks, carotid artery disease, symptomatic peripheral artery disease, abdominal aortic aneurysm). The presence or absence of diabetes, by medication history of glucose evaluation, adds to the list of CHD equivalents.

Continuing the development of the database involves evaluating the patient's blood pressure history. Essential hypertension has long been a predictor of CVD and is a significant player in the development of the Framingham CHD risk calculator. Of importance, the systolic blood pressure value, not the diastolic value, is most predictive of CHD risk. This is true whether the patient is on antihypertensive therapy or treatment naïve. The pharmacist should continue documenting blood pressure readings as the treatment plan evolves because improvements in blood pressure control can modulate CHD risk over time.

Smoking status requires a careful assessment. The number of cigarettes consumed should be assessed and reassessed at each pharmaceutical care contact. Smoking status imparts significant cardiovascular risk. The current guidelines count smoking as a positive risk factor if the patient has smoked within the past month. Because smoking cessation is difficult for some individuals, the status of cigarette or other tobacco consumption should be constantly reassessed. Smoking has been documented to lower HDL-C, which may be one of the reasons that cigarette smokers have increased rates of CHD. Note that the significance of smoking on the Framingham risk score for primary prevention declines with increasing age. Nevertheless, with current pharmacotherapy, smoking cessation is in reach of many adults.

As noted previously, other risk factors should be considered when developing the treatment plan.

Exercise and BMI, including WC in those who have symptoms suggestive of metabolic syndrome, require careful assessment. One of the criteria for metabolic syndrome is a WC greater than 40 inches in men and 35 inches in women. What the patient does for physical activity requires careful documentation. The use of clinical pedometry, whereby well-engineered but affordable pedometers[6] are prescribed and weekly step-counts are recorded and followed according to that required for cardiometabolic risk reduction, is an example. As noted elsewhere in this publication, exercise is essential for improving diabetic indices, raising HDL-C, lowering elevated triglycerides, and maintaining a healthy body weight. According to government guidelines, Americans should engage in 30 minutes of moderate-intensity physical activity 5 or more days of the week.[7] The current physical activity guidelines for managing overall dyslipidemia are published by the American College of Sports Medicine, which now recommends physical activity 5 or more days per week for 40–60 minutes/day at 40–70% of aerobic capacity.[7] Constant reminders of physical inactivity can be one of the key roles of the pharmacist as a concerned health care practitioner.

The pharmacist should pay particularly close attention to the diet of the patient. Although, admittedly, pharmacists are not dietitians, as involved health care individuals, they can often make great strides in keeping patients on track. Overconsumption of high carbohydrate–containing foods, foods high in saturated fats, and foods high in sodium content is an important contributor to cardiovascular risk. The American diet remains one of the more important characters in the high rate of CVD and requires constant reminders to maintain moderation in the intake of offending foods.

In documenting other information, the pharmacist should pay close attention to other thrombotic disorders. Although not well defined scientifically, patients receiving therapy for thrombotic disease (e.g., venous thromboembolism, arterial clotting disorders) may be at risk of CHD. The pharmacist may casually note this information in the database, but there are no current recommendations for other therapies based on this knowledge.

Finally, in the construct of the pharmacist's database, adverse drug reactions require attention. The nature, temporal sequence, symptoms, and plausibility of adverse drug reactions or drug allergies will be essential in helping the health care provider determine the appropriate regimen. The adverse drug reaction database should be updated at each pharmaceutical care encounter.

PATIENT EDUCATION

The pharmacist can play a key role in educating patients with respect to their need for drugs and other interventions altering the various lipoproteins. Both patients and their significant other(s) should be present at the session. Group sessions are often very successful because they tend to build other support networks. Topics discussed at these education sessions are (1) the pathophysiology of cardiometabolic disease and its relation to other vascular disease, (2) TLCs, and (3) the drugs used for therapy. Useful resources are the National Lipid Association, the Preventive Cardiovascular Nurse's Association, the AHA, the NHLBI, and the American Dietetic Association.

PHYSICAL ASSESSMENT

As noted previously, assessment of the systolic and diastolic pressure is important in the development of the pharmacist's database. Many pharmacies are equipped with good blood pressure monitors. The NHLBI[8] provides tips for patients and professionals to consider when obtaining measurements.

Cigarette smoking 30 minutes before measurements are obtained can affect blood pressure. In addition, vasoreactions caused by a full bladder can affect blood pressure measurements. The patient should be sitting quietly for 5 minutes before blood pressure is taken and should not talk during measurement. Obstructive clothing and improper arm position can cause inaccurate measurements. The final number should be the average of two successive measurements.

Monitoring the patient's weight and calculating a BMI will require a scale and a calculation table, which are available from multiple sources. Most practitioners allow 5 lb for clothing in calculating actual body weight. Weight should be continually monitored because it usually changes with life events, changing seasons, and activity level.

A careful evaluation of the skin and cutaneous structures should be attempted. The presence of lipomas may indicate a lipid disorder. Lipomas are usually benign and not treated medically or surgically. Lipomas are common, benign, mesenchymal tumors. They may develop in virtually all organs of the body. Noncancerous lesion features include soft, fluctuant feel; lobulation; and the free mobility of overlying skin. Malignant lesions will have a slippery feeling as opposed to the more formed non-malignant structure. Arcus senilis, a peripheral corneal opacity, may be indicative of advanced dyslipidemia. Hypertrophied Achilles tendons (xanthomas) are often suggestive of a familial dyslipidemia.

The pharmacist should be able to assess the carotid arteries for pulses and the presence or absence of a jugular venous pressure. An elevated jugular venous pressure may be suggestive of CVD, including hypertension or heart failure. Peripheral pulses, including pedal pulses, will suggest the presence or absence of other vascular diseases.

The pharmacist should also assess the fifth vital sign, the patient's pain level. Several scales, both visual and analog, are available for assessment. The same scale should be used at each visit.

LABORATORY VALUES

Much of the lipid treatment plan is based on laboratory values. It is generally useful to create a grid or graph over time, plotting changing LDL-C, HDL-C, non-HDL, and triglyceride values together with adjustments. For patients with triglycerides higher than 200 mg/dL, apoB is perhaps a better marker of risk and target of therapy according to a growing consensus of clinical lipidologists.[9] Total cholesterol and ratio values are no longer targets of therapy, but they may be provided for a more complete database. The pharmacist should pay particular attention to whether or not the patient was fasting for 12 hours before the test because triglyceride values are sensitive to postprandial effects. The LDL-C, if calculated, will significantly underepresent the true LDL-C burden when triglycerides become significantly elevated (e.g., more than 400 mg/dL), although this disconnect begins with triglycerides higher than 200 mg/dL. Many institutions are now providing direct LDL-C measurements, including LDL-C particle concentration measured by NMR (LipoScience), to preclude this influence. There are instances in which advanced lipoprotein, apoprotein, and other biomarker assays are important (e.g., apoB, Lp(a), LDL-C particle concentration, high-sensitivity CRP, homocysteine), and the pharmacist should decide beforehand which patient cases will require these tests.

Fasting glucoses or fingerstick values should be recorded in the database to assess diabetic management. A baseline thyroid-stimulating hormone and SCr should be recorded to determine secondary causes of dyslipidemia. Other safety values such as CPK or CK (creatinine phosphokinase or creatinine kinase) and ALT (serum glutamate pyruvate transaminase; alanine transaminase) should be noted at the beginning of the treatment plan and may become important when assessing the adverse effects of drug therapy. Currently, it is not believed that CPK and ALT require routine monitoring during drug therapy.[10] Laboratory indications for fatty liver (nonalcoholic steatohepatitis) are important.

ADHERENCE

The ATP III has identified the role of the pharmacist on the medication management team as an important contributor to enhancing adherence to drugs. Nonadherence to the TLCs as well as medication management contributes to poor clinical outcomes. Research has confirmed that there is a continually diminishing level of adherence, with at least 25% of patients in all groups discontinuing drug therapy by 6 months. In the United States, statin therapy adherence is very likely closer to 50% at 1 year, partially because Americans tend to have a higher copayment for drugs or may lack insurance that covers drugs.[11]

Regardless of the method used to improve compliance, the pharmacist should assess medication adherence at each patient care encounter. If the patient is thought to be nonadherent, ascertain whether it is because of an adverse drug reaction, health beliefs, published information, misunderstood instructions, or missed refills. This is particularly important if the nonadherence was not communicated to the patient's health care provider who authorized a dose increase or change in drug. Some common reasons for nonadherence include missing evening doses of statins that are prescribed for bedtime administration. Often, changing the administration time to a nonstandard hour can improve medication consumption and clinical outcomes. Medication adverse effects cause patients to stop taking pharmaceuticals or neutraceuticals used for dyslipidemia, and patients often wait until their next appointment to communicate these issues to their provider.

The pharmacist should offer medication organizers or medication calendars to the patient to assist in remembering when and how to take drugs for dyslipidemia. Table 2 lists overall strategies for improving patient compliance with lipid-lowering therapies.

Finally, the pharmacist-provider should use strategies commensurate with current behavioral science. A combination of strategies (e.g., behavioral counseling, educational approaches, supportive techniques) is strongly recommended, as is targeting the multiple levels of adherence.[12] Beginning with the patient, the pharmacist needs to determine not only whether the patient is ready to make a change and is confident about implementing the treatment, but also whether the patient has the knowledge, skills, and resources to start the plan. Perhaps the most evidence-based behavioral strategy to accomplish the empowering of patients to make healthy decisions and improving compliance is using the process of motivational interviewing (MI) during patient visits. Motivational interviewing can effectively help clients modify addictive behaviors such as unhealthy eating habits and sedentary lifestyle and can identify two or three behaviors that the client is willing to change. Motivational interviewing also helps the client acknowledge the role of diet and physical activity as components of

TABLE 2. METHODS TO IMPROVE ADHERENCE WITH DYSLIPIDEMIA THERAPY

1. The provider should maintain open communication with the patient that permits a discussion of the many factors that may influence a patient's adherence.
2. If cost is a factor and the patient cannot afford lipid-lowering therapy drugs, then the provider should try alternative strategies, and dietary therapy should be emphasized. Even dietary therapy may need to be introduced gradually, with regular checkups to determine how the patient is progressing in implementing dietary changes.
3. The provider must clearly demonstrate the ability to teach and convey the risk reduction value of the prescribed therapy. For example, with lifestyle modifications, if providers value and engage in healthy dietary and physical activity themselves, they become better advocates and can articulate more realistic lifestyle modifications for the patient.
4. The provider should empower the patient to make healthy decisions and improve adherence by using the process of motivational interviewing during patient visits.

medical therapy. Key tenets of MI are *expressing empathy, reducing ambivalence and developing discrepancy, facilitating self-motivational statements, avoiding or rolling with resistance, and using microcounseling skills*. An excellent and authoritative resource for becoming familiar with this process is at motivationalinterviewing.org.[13]

THE PHARMACEUTICAL CARE TREATMENT PLAN

Documentation of the pharmacist's treatment plan is essential to consider opportunities for reimbursement as well as for the long-term treatment of the patient. In addition to the database elements previously described, the pharmacist must define, in a problem-oriented fashion, the patient's treatment goals, objective information, and strategies used to produce desired clinical outcomes. Recently, the AMA's Current Procedural Technology editorial panel created new codes to allow coding for pharmacist–patient interactions. The reimbursement for pharmaceutical care of patients is slowly becoming available. To begin the process, all pharmacists seeking reimbursement must apply for a National Provider Identification number. More information is available at nppes.cms.hhs.gov. Pharmacists who seek reimbursement for medication management are currently limited to large organizations, where they have support from the business office.

Typical Current Procedural Terminology codes for pharmacists' medication management visits are 99212 or 99213 for initial visits and 99211 for follow-up visits.[14,15] Medicare's "incident-to" guidelines perhaps provide the best authoritative guidance on non-physician billing when providing such services in outpatient physician offices. Most states require the non-physician provider to be a mid-level provider (e.g., nurse practitioner, physician assistant) to bill above a 99211 level. The pharmacist also must know disease state coding terminology such as ICD-9 code 272 for disorders of lipid metabolism, 272.1 for pure hypertriglyceridemia, 272.2 for mixed hyperlipidemia, and 272.4 for other and unspecified hyperlipidemia.[16] The treatment plan must contain elements that reference the coding for these medication management visits.

PRACTICE MODELS

The Department of Veterans Affairs documents many instances in which pharmacists[17] and other health care practitioners have established lipid clinics. In the model established in 1988, medical, cardiology, pharmacist, and dietetic disciplines collaborated to define patient treatment plans. The clinic still functions, although somewhat modified, 19 years later. Based on per protocol recommendations, the clinic-reported outcomes that are related to aggressive patient management have shown a decrease in cardiac-related admissions[18] as well as cost-effective use of drugs.

A newer Veterans Affairs clinic functioned based solely on the pharmacist's recommendations without a predetermined protocol. Recommendations for patient management were approved by an attending physician, who prescribed the drugs. Outcomes were reported for 284 patients after 3 years. Compliance with lipid-lowering agents ranged from 43% to 100%. The clinic as originally established no longer exists.[19]

The benefit of the pharmacist versus a usual care model was evident in a recent publication. A clinical pharmacy group's lipid recommendations were compared with the outcome of lipid recommendations in a primary care Veterans Affairs setting. Records (47) from the clinical pharmacist group were compared with a usual care (41) group. The pharmacist's management of patients showed an average of 18.5% LDL-C lowering compared with 6.5% in the usual care group. It was also noted that more clinical pharmacy visits were associated with a greater magnitude of LDL-C change.[20]

Community pharmacists have shown the ability to provide clinical impact. Project Impact was a controlled experiment in six community pharmacies. Using a fingerstick cholesterol meter, pharmacists provided both counseling and medication recommendations to providers. Five hundred twelve initial drug therapy problems were reported for 116 patients during the study. Both lipid and non-lipid management strategies were addressed by the participating pharmacists. There were 354 (69.1%) drug therapy problems associated with hyperlipidemia and 158 (30.9%) drug therapy problems associated with other conditions. Drug adherence was also addressed. The project demonstrated that non-institutional pharmacists could play an integral role in managing patient therapy.[21]

In a group practice clinic located in Hawaii, the outcomes of two cohorts were compared. Each group practice had five attending physicians, five third-year medical students, four second-year medical students, and six medical interns. In the intervention group, a pharmacist routinely provided advice on appropriate lipid management and interacted with patients to assist in treatment plan implementation. Forty-seven patients were managed in both the pharmacist intervention and usual care control groups, and the lipid results were compared. During the six-month study, the pharmacist group made 186 therapy recommendations with a 90% acceptance rate. The total cholesterol values dropped (44 ± 47 mg/dL vs. 13 ± 51 mg/dL in the usual care group). There was an additional cost decrease in the pharmacist group ($11/month) versus a cost increase in the control population ($4/month).[22]

The pharmacist has improved patient satisfaction, in addition to lipid outcomes. The Cleveland Veterans Affairs Medical Center studied patients who attended that medical center's lipid clinic. A total of 224 patients and 109 providers who interacted with the lipid clinic were chosen to participate in a survey. The patient response to the survey was 47% of the total participants. Patient satisfaction with pharmacist care, patient education materials, and education about cholesterol, diet, and exercise was extremely high. The provider response was also positive. Most providers responded that they were strongly or somewhat satisfied with the care provided by the pharmacist. Of additional note, the providers found the progress notes written by the pharmacist, as well as the pharmaceutical care provided, very helpful. Positive lipid-lowering outcomes were noted in the review, but no control population was established.[23]

Finally, the National Lipid Association provides a wide array of lipid clinic development articles, CDs, and courses (www.lipid.org) for its members.

SUMMARY

The pharmacist plays an important role on the health care team. As a front-line team member, the pharmacist can positively affect adherence and disease state outcomes. There are many resources, including training programs, to enhance the pharmacist's knowledge of clinical lipidology and practical therapeutic lifestyle strategies to help manage dyslipidemia. It becomes rewarding to know that by instituting a systematic approach to lifestyle and pharmacotherapy, the pharmacist can improve the quality and possibly the length of a patient's life.

REFERENCES

1. Hepler CD, Strand LM. Opportunities and responsibilities in pharmaceutical care. Am J Pharm Educ 1989;53:7S–15S.
2. National Cholesterol Education Program. Available at www.nhlbi.nih.gov/guidelines/cholesterol/atp3full.pdf. Accessed July 26, 2008.
3. National Cholesterol Education Program. Risk assessment tool. Available at http://hp2010.nhlbihin.net/atpiii/calculator.asp?usertype=prof. Accessed July 29, 2008.
4. Davignon J, Dufour R. Lipid Disorders: An Atlas of Investigation and Management. Boca Raton, FL: Clinical Publishing, 2007.
5. Accreditation Council for Clinical Lipidology. Eligibility criteria. Available at www.lipidspecialist.org/criteria.php. Accessed June 30, 2008.
6. Available.www.accu-split.com and http://www.pedometersusa.com/pedometers-all.html. Accessed July 28, 2008.
7. American College of Sports Medicine. ACSM's Guidelines for Exercise Testing and Prescription, 7th ed. Philadelphia: Lippincott Williams & Wilkins, 2006.
8. National Institutes of Health. Available at www.nhlbi.nih.gov/guidelines/hypertension/phycard.pdf. Accessed July 23, 2008..
9. Barter PJ, Ballantyne CM, Carmena R, et al. Apo B versus cholesterol in estimating cardiovascular risk and in guiding therapy: report of the thirty-person/ten country panel. J Intern Med 2006;259:247–58.
10. McKenney JM. Report of the National Lipid Association's Statin Safety Task Force. Am J Cardiol 2006;97(suppl 1):S1–S98.
11. Compliance Action Program. Available at www.americanheart.org/presenter.jhtml?identifier=1657. Accessed August 2007.
12. Fletcher B, Berra K, Ades P, et al. Managing abnormal blood lipids. Circulation 2005;112:3184–209.
13. Motivational interviewing. Available at www.motivationalinterviewing.org. Accessed June 30, 2008.
14. Jarrett AT. Understanding basic concepts and strategies for obtaining pharmaceutical reimbursement. Am J Health Syst Pharm 2006;63(suppl 7):S7–9.
15. Nutescu EA, Klotz RS. Basic terminology in obtaining reimbursement for pharmacists' cognitive services. Am J Health Syst Pharm 2007;64:186–92.
16. Kuo GM, Buckley TE, Fitzsimmons DS, Steinbauer JR. Collaborative drug therapy management services and reimbursement in a family medicine clinic. Am J Health Syst Pharm 2004;61:343–54.
17. Kellick K, Burns, K, McAndrew E. Diet and drug therapy in the management of hyperlipidemia. VA Pract 1991;8:71–80.
18. Kellick KA. Outcomes monitoring from a Department of Veterans Affairs Lipid Clinic—focus on fluvastatin. Poster presentation 1994 International Atherosclerosis Symposium, Montreal, CA.
19. Furmaga EM. Pharmacist management of a hyperlipidemia clinic. Am J Health Syst Pharm 1993;50:91–5.

20. Till LT, Voris JC, Horst JB. Assessment of clinical pharmacist management of lipid-lowering therapy in a primary care setting. J Manag Care Pharm 2003;3:269-73.
21. McDonough RP, Doucette WR. Drug therapy management: an empirical report of drug therapy problems, pharmacists' interventions, and results of pharmacists' actions. J Am Pharm Assoc 2003;43:511-8.
22. Bogden PE, Koontz LM, Williamson P, Abbott RD. The physician and pharmacist team. An effective approach to cholesterol reduction. J Gen Intern Med 1997;12:158-64.
23. Collins C, Kramer A, O'Day ME, Low MB. Evaluation of patient and provider satisfaction with a pharmacist-managed lipid clinic in a Veterans Affairs medical center. Am J Health Syst Pharm 2006;63:1723-7.

INDEX

Page numbers followed by *f* or *t* indicate material in figures or tables, respectively.

A

ABI (ankle brachial index), 139–40, 140*f*
ACCESS study, 143
Accreditation Council for Clinical Lipidology, 157–58
ACE inhibitors, 53
Achilles tendon xanthoma, 161
Action to Control Cardiovascular Risk in Diabetes, 53
acute coronary syndrome, 93*t*, 94, 106–9, 108*t*
adenosine triphosphate–binding cassette A1 (ABCA1), 6
adenosine triphosphate–binding cassette G1 (ABCG1), 6, 9
adherence, 162–63, 163*t*
adhesion molecules, in atherogenesis, 2*f*, 3
adipokines, in metabolic syndrome, 45–46
adiponectin: in metabolic syndrome, 45–46; as therapeutic target, in obesity, 55
adipose tissue: as endocrine organ, 45; in metabolic syndrome, 45–46
adult patient, clinical evaluation of, 20–26
Adult Treatment Panel III (ATP III): on clinical guidelines, 84; on diet, 14, 32–36; on metabolic syndrome, 42, 43*t*; on risk factors, 156–57; on therapeutic lifestyle changes, 30–40
advanced lipid tests, 23–26, 26*t*, 141–42
AFCAPS/TexCAPS trial, 85, 86*t*, 122, 123*t*, 142
African Americans, metabolic syndrome in, 44
Aggrastat to Zocor (A to Z) trial, 93*t*, 94, 106
AHA. *See* American Heart Association
AHA/Advanced Cardiovascular Life Support, 106
AHA Atherosclerosis, Hypertension, and Obesity in Youth Committee, 104
Air Force/Texas Coronary Atherosclerosis Prevention Study (AFCAPS/TexCAPS), 85, 86*t*, 122, 123*t*, 142
ALA (α-linoleic acid), 35–36, 36*t*
alcohol consumption, 14, 15*t*, 153
ALLHAT-LLT trial, 87–91, 90*t*
allicin (garlic), 154
alpha-tocopherol (vitamin E), 153
American Board for the Certification of Lipid Specialists, 157–58
American College of Cardiology, clinical guidelines of, 84–85
American College of Sports Medicine, 160
American Dietetic Association, 160

American Heart Association (AHA): on acute coronary syndrome, 106; clinical guidelines of, 84–85; on diet, 32, 36, 80, 143, 152; on metabolic syndrome, 42, 43t, 49–50; patient education resources of, 160; on pediatric patients, 104
amiodarone, and lipid disorders, 15t, 16
angiotensin-converting enzyme (ACE) inhibitors, 53
angiotensin receptor blockers (ARBs), 53
Anglo-Scandinavian Cardiac Outcomes Trial—Lipid Lowering Arm (ASCOT-LLA), 50–52, 85, 86t
ankle brachial index (ABI), 139–40, 140f
anorexia nervosa, 14, 15t
antacids, interaction with ezetimibe, 71
Antara. *See* fenofibrate
anticonvulsants, and lipid disorders, 15t, 16
Antihypertensive and Lipid-Lowering Treatment to Prevent Heart Attack Trial (ALLHAT-LLT), 87–91, 90t
antipsychotics, and lipid disorders, 15t, 16
antiretroviral therapy: and HIV-associated dyslipidemia, 111–20, 117t; and statin use, 116, 118t; switching of agents, 114–16, 115t
apheresis, in pregnant patients, 101t
apolipoprotein(s), 5–6, 5t, 6f
apolipoprotein A-I, 6
apolipoprotein A-II, 6
apolipoprotein B: in atherogenesis, 1, 3; as risk marker, 143
apolipoprotein B-48, 5
apolipoprotein B-100, 5–6; familial defective, 11, 11t; Lp(a) and, 14
apolipoprotein CII: as cofactor, 5; familial deficiency of, 12
apolipoprotein E, 5–6; isoforms of, 12–13; mutation of, disorder with, 12–13
apoprotein AI deficiency, 13
apoptosis, in atherogenesis, 4
appetite suppressants, 54
arachidonic acid, 35
ARBs (angiotensin receptor blockers), 53
arcus cornea (arcus senilis), 10, 161
ascorbic acid (vitamin C), 153
ASCOT-LLA trial, 50–52, 85, 86t
Asian patients, 91, 93t
aspirin therapy, in metabolic syndrome, 55
ASTEROID trial, 94, 95t
atazanavir: and dyslipidemia, 111, 114–16, 115t; and statin use, 118t
atherogenesis, 1–4, 2f
atheroma burden, as surrogate marker, 94, 95t
atherosclerotic vascular disease, pathophysiology of, 1–16
Atkins diet, 38, 39t
atorvastatin, 60–67; for acute coronary syndrome, 93t, 94, 106, 108t, 109; administration of, 67, 144; adverse effects of, 64–66, 145; and apolipoprotein B,

143; and atheroma burden, 95t; in diabetic patients, 50, 91, 92t; dosing of, 63f, 64t, 67, 144, 149t; drug interactions of, 66–67, 68t–69t, 118t, 145; effects on lipid parameters, 62, 62t, 149t; in elderly patients, 123t, 124; for HIV-associated dyslipidemia, 117t, 118t; mechanism of action, 61–62, 61f; monitoring with, 146–47; in pediatric patients, 107t; pharmacokinetics/pharmacodynamics of, 62–64, 63t; in pregnancy, avoidance of, 101t, 102; primary prevention trial of, 86t; renal considerations with, 112t; secondary prevention trial of, 88t, 89t; for transplant-related dyslipidemia, 121t
Atorvastatin Comparative Cholesterol Efficacy and Safety Study (ACCESS), 143
A to Z trial, 93t, 94, 106
ATP III. *See* Adult Treatment Panel III
atypical antipsychotics, and lipid disorders, 15t, 16
AURORA study, 91
azathioprine, and dyslipidemia, 119t, 120
azole antifungals, statin interaction with, 67, 69t

B

BABRs. *See* bile acid sequestrants
BASs. *See* bile acid sequestrants
β-blockers, and lipid disorders, 14, 15t
beta-carotene, 153
bile acid binding resins. *See* bile acid sequestrants
bile acid sequestrants (BASs), 60, 71–73, 148–49; administration of, 73, 148; adverse effects of, 72, 148; dosing of, 73, 73t; drug interactions of, 72, 148; effect on lipid parameters, 71t; in elderly patients, 124; intestinal action of, 147–48; mechanism of action, 72; monitoring with, 149; in pediatric patients, 105, 107t, 148; pharmacokinetics/pharmacodynamics of, 72; in pregnant patients, 101t, 102; for renal-associated dyslipidemia, 110t; renal considerations with, 112t; for transplant-associated dyslipidemia, 121
blood pressure: assessment, in lipid clinic service, 159, 160; in metabolic syndrome, 42–53, 43t
board certification, in clinical lipidology, 157–58
body mass index (BMI), 38t, 158, 159, 161. *See also* obesity
breastfeeding, dyslipidemia treatment in, 102
bupropion, for weight loss, 54
burns, and lipid abnormalities, 15t

C

CAD (coronary artery disease). *See* cardiovascular disease
calcium channel blockers, statin interaction with, 67, 69t
cannabinoid receptors, in obesity, 55

cardiometabolic risk, 42. *See also* metabolic syndrome
cardiovascular disease (CVD): emerging risk markers for, 26, 27t–28t;
 lifestyle modification and, 30–40; metabolic syndrome and, 42–55;
 primary prevention of, 30; risk factor clusters for, 42;
 risk stratification for, 22–26, 22f, 24t
CARDS trial, 50, 91, 92t
CARE trial, 85–87, 88t, 122, 123t
carotid arteries, evaluation of, 161
carotid B-mode ultrasound, 140
cell adhesion molecules, in atherogenesis, 2f, 3
certification, in clinical lipidology, 157–58
CETP. *See* cholesterol ester transfer protein
CHD (coronary heart disease). *See* cardiovascular disease
chemokines, in atherogenesis, 3
chemotactic factors, in atherogenesis, 3
children, dyslipidemia in, 103–6; lifestyle modifications for, 105;
 pharmacotherapy for, 105–6, 105t, 107t–108t, 146–48;
 risk stratification for, 103, 104t; treatment recommendations for, 103–6
cholestasis, 16
cholesterol, 4–5; biological functions of, 4; biosynthesis of, 4; carrier proteins
 for, 5 (*See also* lipoprotein); dietary, 4, 14, 15t, 33–34; elevated levels of. *See*
 Hypercholesterolemia; endogenous, metabolism of, 8, 8f; exogenous, absorption
 and transport of, 6–7, 7f; reverse transport of, 9, 9f; serum levels of, 5, 20–21, 21t
cholesterol absorption inhibitors, 60, 67–71, 149; adverse effects of, 70; drug
 interactions of, 71, 75, 121t; effect on lipid parameters, 70t; in elderly patients,
 124; with fenofibrate, 70, 70t; for HIV-associated dyslipidemia, 116, 117t;
 mechanism of action, 67–70; in pediatric patients, 106, 107t; pharmacokinetics/
 pharmacodynamics of, 70; in pregnant patients, 101t, 102; for renal-associated
 dyslipidemia, 110t; renal considerations with, 112t; for transplant-associated
 dyslipidemia, 121, 121t
Cholesterol and Recurrent Events (CARE) trial, 85–87, 88t, 122, 123t
cholesterol ester transfer protein (CETP): deficiency of, 14;
 insulin resistance and, 48, 49f; in lipid metabolism, 9
cholestyramine, 71–73, 148–49; administration of, 73, 148; adverse effects of, 72, 148;
 dosing of, 73, 73t; drug interactions of, 68t, 71, 72, 74–76, 78, 148;
 intestinal action of, 147–48; mechanism of action, 72; monitoring with, 149;
 in pediatric patients, 107t, 148; pharmacokinetics/pharmacodynamics of, 72;
 in pregnant patients, 101t, 102; renal considerations with, 112t
chronic kidney disease: dosing considerations in, 112t–113t;
 dyslipidemia in, 15t, 16, 109–11; lipid-lowering therapy in, 110–11, 111t;
 screening/evaluation in, 110
chylomicron(s), physical characteristics of, 5, 5t
chylomicron remnants, 7, 7f
cigarette smoking. *See* smoking
cimetidine, interaction with ezetimibe, 71

circadian rhythm, in cholesterol synthesis, 4
CKD. *See* chronic kidney disease
Cleveland Veterans Affairs Medical Center, 165
clinical evaluation, of adult patient, 20–26
clinical guidelines, 84–85
clinical lipidology, credentials in, 157–58
clinical practice pearls, 139–54
Clinical Quality Improvement Network Investigators, 124
clinical trials, 84–94; in elderly patients, 91, 122–24, 123*t*; mixed-population, 87–91, 90*t*; primary prevention, 85, 86*t*; secondary prevention, 85–87, 88*t*–89*t*; in special populations, 91–94, 92*t*–93*t*; surrogate markers assessed in, 94, 95*t*
clinic service, 156–65. *See also* lipid clinic service
coenzyme Q10, 154
colesevelam, 71–73, 148–49; administration of, 73, 148; adverse effects of, 72, 148; dosing of, 73, 73*t*; drug interactions of, 71, 72, 148; intestinal action of, 147–48; mechanism of action, 72; monitoring with, 149; in pediatric patients, 107*t*, 148; pharmacokinetics/pharmacodynamics of, 72; in pregnant patients, 101*t*, 102; renal considerations with, 112*t*
colestipol, 71–73, 148–49; administration of, 73, 148; adverse effects of, 72, 148; dosing of, 73, 73*t*; drug interactions of, 68*t*, 71, 72, 74–75, 78, 148; intestinal action of, 147–48; mechanism of action, 72; monitoring with, 149; in pediatric patients, 107*t*, 148; pharmacokinetics/pharmacodynamics of, 72; in pregnant patients, 101*t*, 102; renal considerations with, 112*t*
Collaborative Atorvastatin Diabetes Study (CARDS), 50, 91, 92*t*
combined hyperlipidemia, 12–13
complex lipid disorders, 157, 157*t*
compliance, 162–63, 163*t*
computed tomography, electron beam, 141
constipation, bile acid sequestrants and, 72
corneal changes, 10, 13, 161
coronary artery disease (CAD). *See* cardiovascular disease
coronary heart disease. *See* cardiovascular disease
CPT coding, 163–64
C-reactive protein: high-sensitivity, as risk marker, 27*t*; in metabolic syndrome, 42, 45, 54, 55; in obesity, 54, 55
credentials, in clinical lipidology, 157–58
CRP. *See* C-reactive protein
Current Procedural Terminology, 163–64
CVD. *See* cardiovascular disease
cyclooxygenase 1 (COX-1), 3
cyclooxygenase 2 (COX-2), 3
cyclosporine: interaction with ezetimibe, 71, 121*t*; interaction with fenofibrate, 121*t*; interaction with statins, 67, 68*t*, 120–21, 121*t*; lipid abnormalities with, 15*t*, 16, 119*t*, 120
cytokines, in metabolic syndrome, 44–46

D

4D's, 14–16, 15t
4D (clinical trial), 91, 92t
darunavir, and statin use, 118t
database, for lipid clinic service, 158–59
Data Collection on Adverse Events of Anti-HIV Drugs Study Group, 114
delavirdine, and statin use, 117t
DHA (docosahexaenoic acid), 35, 37t, 79–80, 152
diabetes mellitus: drugs for, and lipid disorders, 16; nicotinic acid and, 78, 151–52
diabetes mellitus, type 2: cardiovascular risk in, management of, 50–53, 91, 92t; as cause of lipid disorders, 15t, 16; in metabolic syndrome, 42–55; obesity and, 38; prevention, in prediabetes, 52, 54
Diabetes Prevention Program Research Group, 54
diastolic blood pressure, 159, 160
diet: as cause of lipid disorders, 14, 15t; cholesterol in, 4; for elderly patients, 124; for hypertension, 53; for metabolic syndrome, 50; pharmacist role in, 160; supplements in, 152–54; therapeutic changes in, 30, 31–36, 33t, 50, 124; for weight loss, 29t, 38–39, 54
Dietary Approaches to Stop Hypertension, 53
diethylpropion, 54
digoxin, bile acid sequestrants with, 148
diltiazem, statin interaction with, 68t
diuretics, and lipid disorders, 14, 15t
docosahexaenoic acid (DHA), 35, 37t, 79–80, 152
drug-induced lipid disorders, 14–15, 15t
drug therapy. *See* pharmacotherapy
dysbetalipoproteinemia, familial, type III, 12–13
dyslipidemia. *See* lipid disorders; *specific types*

E

education, patient, 160
efavirenz: and dyslipidemia, 114–16, 115t; and statin use, 117t
eicosapentaenoic acid (EPA), 35, 37t, 79–80, 152
elderly, dyslipidemia in, 122–24; clinical trials on, 91, 122–24, 123t; 1%/2% rule and, 122; pharmacotherapy for, 91, 92t, 122–24, 123t
electron beam computed tomography (EBCT), 141
emerging risk markers, 26, 27t–28t, 50, 139–43
endocannabinoid system, in obesity, 54–55
EPA (eicosapentaenoic acid), 35, 37t, 79–80, 152
eructation, omega-3 fatty acids and, 79
eruptive xanthoma, 12

INDEX

erythromycin, statin interaction with, 67, 68t
E-selectin, 3
estrogens: exogenous, and lipid disorders, 14–16, 15t; in pregnancy, and lipid disorders, 100
ethnicity, and metabolic syndrome, 43–44
evidence-based medicine, 84–85. *See also specific trials and studies*
exercise, 30, 36–37, 124, 159–60
exercise tolerance test, 140–41
expert consensus documents, 84–85
ezetimibe, 67–71, 149; adverse effects of, 70; drug interactions of, 71, 75, 121t; effect on lipid parameters, 70t; in elderly patients, 124; with fenofibrate, 70, 70t; for HIV-associated dyslipidemia, 116, 117t; mechanism of action, 67–70; in pediatric patients, 106, 107t; pharmacokinetics/pharmacodynamics of, 70; in pregnant patients, 101t, 102; for renal-associated dyslipidemia, 110t; renal considerations with, 112t; for transplant-associated dyslipidemia, 121, 121t

F

familial apolipoprotein CII deficiency, 12
familial combined hyperlipidemia, 11t, 12
familial defective apolipoprotein B-100, 11, 11t
familial dysbetalipoproteinemia (type III), 12–13
familial dyslipidemias, 10–14
familial heterozygous hyperlipidemia, 105
familial HL deficiency, 13
familial homozygous hyperlipidemia, 102
familial hyperalphalipoproteinemia, 14
familial hypercholesterolemia, 10, 11t
familial hypoalphalipoproteinemia, 13
familial LPL deficiency, 12
fasting lipid profile, 20–21
fat(s): dietary, 14, 15t, 33–34, 34t, 50; saturated, 33, 34t, 50; trans, 33–34, 50; unsaturated, 33–34, 34t, 50
fatty acid(s): dietary, 33–36, 36t, 37t; free, in metabolic syndrome, 45; nonesterified, 45–48; omega-3. *See* omega-3 fatty acids; omega-6, 33, 35–36, 36t; in triglycerides, 4
fenofibrate, 73–75, 147–48; administration of, 75–76; adverse effects of, 74; in diabetic patients, 53, 91; dosing of, 75–76, 147; drug interactions of, 71, 74–75, 121t; in elderly patients, 124; with ezetimibe, 70, 70t; mechanism of action, 74; in metabolic syndrome, 53; in pediatric patients, 107t; pharmacokinetics/ pharmacodynamics of, 74; in pregnant patients, 101t, 103; renal considerations with, 113t; for transplant-related dyslipidemia, 121–22, 121t
Fenofibrate Intervention and Event Lowering in Diabetes (FIELD), 91, 92t
fiber, dietary, 35
fibrates. *See* fibric acid derivatives

fibric acid derivatives, 53, 60, 73–75, 147; administration of, 75–76; adverse effects of, 74; in diabetic patients, 53, 91; dosing of, 75–76; drug interactions of, 74–75; in elderly patients, 124, 124t; for HIV-associated dyslipidemia, 116–20, 117t; mechanism of action, 74; in metabolic syndrome, 53; National Lipid Association on, 75, 76t, 85; in pediatric patients, 107t; pharmacokinetics/pharmacodynamics of, 74; in pregnant patients, 101t, 103; for renal-associated dyslipidemia, 110t, 111; renal considerations with, 75, 113t; safe prescribing of, 75, 76t; statins with, 147; for transplant-related dyslipidemia, 121–22

fibrinogen: as risk marker, 27t; as therapeutic target, 55

FIELD trial, 91, 92t

Finnish Diabetes Prevention Study, 52, 54

fish-eye disease, 13

fish/fish oil, 33, 34t, 36, 37t, 152. *See also* omega-3 fatty acids

flaxseed/flaxseed oil, 36t, 152–53

fluoxetine, for weight loss, 54

flushing, nicotinic acid and, 77

fluvastatin, 60–67; administration of, 67, 144; adverse effects of, 64–66, 145; and apolipoprotein B, 143; dosing of, 63f, 64t, 67, 144, 149t; drug interactions of, 66–67, 68t–69t, 118t, 145; effects on lipid parameters, 62t, 149t; extended-release, 63; for HIV-associated dyslipidemia, 117t, 118t; monitoring with, 146–47; in pediatric patients, 107t; pharmacokinetics/pharmacodynamics of, 62–64, 63t; in pregnancy, avoidance of, 101t, 102; renal considerations with, 112t; for transplant-related dyslipidemia, 121t

foam cells: in atherogenesis, 3; in metabolic syndrome, 45–46

folic acid, 143

fosamprenavir, and statin use, 118t

4D's of secondary lipid disorders, 14–16, 15t

4D trial, 91, 92t

4S trial, 85–87, 88t, 122, 123t

Framingham Heart Study, 143

Framingham Risk Score, 22, 24t, 84, 159

Frederickson classification, 10, 11t

free fatty acids, in metabolic syndrome, 45

Friedewald equation, 5

G

garlic, 153–54

gel electrophoresis, segmented gradient, 141

gemfibrozil, 73–75, 147–48; administration of, 75–76; adverse effects of, 74; dosing of, 75–76, 147; drug interactions of, 67, 68t, 74–75; in elderly patients, 123t; mechanism of action, 74; in metabolic syndrome, 53; in pediatric patients, 107t; pharmacokinetics/pharmacodynamics of, 74; in pregnant patients, 101t, 103; for renal-associated dyslipidemia, 111; renal considerations with, 113t; for transplant-related dyslipidemia, 121–22

gender, and metabolic syndrome, 43–44
general lipid clinic services, 157
genetics, 10
German Diabetes and Dialysis Study (4D), 91, 92t
glipizide, interaction with ezetimibe, 71
glucocorticoids, exogenous, and lipid disorders, 14–16, 15t
glucose levels: control of, 53; in metabolic syndrome, 43t, 46; monitoring of, 161–62
glycogen storage disease, 15t
"good" cholesterol, 33. *See also* high-density lipoprotein
grapefruit juice, statin interaction with, 67, 69t, 145
group sessions, patient, 160

H

HAART. *See* highly active antiretroviral therapy
Hawaii, lipid clinic model in, 164–65
HDL. *See* high-density lipoprotein
Health Professionals Follow-up Study, 46
heart disease. *See* cardiovascular disease
Heart Protection Study (HPS), 87–91, 90t, 123t
hepatic lipases (HLs): familial deficiency of, 13; insulin resistance and, 48;
  in lipid metabolism, 8
hepatitis, statins and, 146
Hepler and Strand, pharmaceutical care defined by, 156
heterozygous familial hyperlipidemia, 105
high-density lipoprotein (HDL): advanced tests for, 23–26, 26t;
  apolipoproteins of, 6; diet and, 31–36, 54; insulin resistance and, 46–48, 49f;
  metabolism of, disorders of, 13–14; pharmacotherapy and, 52–53, 60–80 (*See also specific drugs*); physical characteristics of, 5, 5t; in pregnant patients, 100;
  serum levels of, 20–21, 21t; transport role of, 9, 9f; treatment goals for, 23
high-density lipoprotein 2 (HDL2), 9
high-density lipoprotein 3 (HDL3), 9
highly active antiretroviral therapy (HAART): and dyslipidemia, 111–20, 117t;
  and statin use, 116, 118t; switching of agents, 114–16, 115t
high-protein diet, 29t, 38–39
high-sensitivity C-reactive protein (hs-CRP), as risk marker, 27t
Hispanics, metabolic syndrome in, 44
HIV-associated dyslipidemia, 111–20; antiretroviral drug switching in, 114–16, 115t;
  pharmacotherapy for, 116–20, 117t, 118t; therapeutic lifestyle changes in, 114, 115t;
  treatment recommendations in, 114–20
HLs. *See* hepatic lipases
HMG-CoA, 4
HMG-CoA reductase, 4, 61, 61f
HMG-CoA reductase inhibitors. *See* statins
homocysteine, as risk marker, 28t, 142–43

homozygous familial hyperlipidemia, 102
HPS trial, 87–91, 90*t*, 123*t*
hs-CRP (high-sensitivity C-reactive protein), as risk marker, 27*t*
human immunodeficiency virus. *See* HIV-associated dyslipidemia
3-hydroxy-3-methyl glutaryl coenzyme A, 4. *See also* statins
hyperalphalipoproteinemia, familial, 14
hypercholesterolemia, 10–12; familial, 10, 11*t*; polygenic, 11–12, 11*t*
hyperglycemia, nicotinic acid and, 78
hyperlipidemia: combined, 12–13; familial combined, 11*t*, 12;
   heterozygous familial, 105; homozygous familial, 102
hypertension: assessment, in lipid clinic service, 159, 160;
   in metabolic syndrome, 42–55; treatment of, 53
hypertriglyceridemia, 12
hypoalphalipoproteinemia, familial, 13
hypothyroidism, and lipid disorders, 15*t*, 16

I

IDEAL trial, 87, 89*t*
IDL. *See* intermediate-density lipoprotein
IDSA/ACTG trial, 114, 115*t*, 117
immunosuppressive therapy, and dyslipidemia, 120–22, 120*t*
Incremental Decrease in End Points through Aggressive Lipid
   Lowering (IDEAL) trial, 87, 89*t*
indinavir, and statin use, 118*t*
Infectious Diseases Society of America and Adult AIDS Clinical Trials Group
   (IDSA/ACTG), 114, 115*t*, 117
inflammation: in atherosclerosis, 1–4, 2*f*; in metabolic syndrome, 44–46
insulin, exogenous, and lipid disorders, 16
insulin resistance: and high-density lipoprotein, 46–48, 49*f*; HIV treatment
   and, 111; in metabolic syndrome, 44–48, 45*f*; nicotinic acid and, 78, 152;
   and triglycerides, 46–48, 48*f*; and very low-density lipoprotein, 46–48, 48*f*
integrase inhibitors, and statin use, 117–19
interferons, and lipid disorders, 15*t*
INTERHEART study, 31
interleukin-6, 45
intermediate-density lipoprotein (IDL): in atherogenesis, 1;
   physical characteristics of, 5, 5*t*
International Diabetes Federation, on metabolic syndrome, 42, 43*t*
interviewing, motivational, 162–63
intestinal absorption and transport, 6–7, 7*f*
intestinal-acting agents, 147–48
intracellular adhesion molecule-1, 3
Isentress (raltegravir), and statin use, 117–19
itraconazole, statin interaction with, 67, 69*t*

J

jugular venous pressure, 161

K

ketoconazole, statin interaction with, 67, 69t
kidney disease: in diabetic patients, 91; dosing considerations in, 112t–113t; fibric acid derivatives and, 75, 113t; lipid abnormalities with, 15t, 16, 109–11; lipid-lowering therapy in, 110–11, 110t; screening/evaluation in, 110; statins and, 65–66, 112t

L

laboratory values, evaluation of, 161–62
LCAT. *See* lecithin cholesterol acyltransferase
LDL. *See* low-density lipoprotein
LDL-related protein (LRP), 7
lecithin cholesterol acyltransferase (LCAT): deficiency of, 13; in lipid metabolism, 9, 9f
leptin, in metabolic syndrome, 45
levothyroxine, bile acid sequestrants with, 72, 148
lifestyle modification. *See* therapeutic lifestyle changes
linoleic acid, 35–36, 36t
lipases, hepatic: familial deficiency of, 13; insulin resistance and, 48; in lipid metabolism, 8
lipemia retinalis, 12
lipid(s), 4–5; endogenous, metabolism of, 8, 8f; exogenous, absorption and transport of, 6–7, 7f; synthesis of, 4–5; transport of, 5–9
lipid clinic service, 156–65; CPT coding in, 163–64; definition of, 156; entry criteria for, 158; general, 157; laboratory values in, 161–62; patient education in, 160; patient evaluation in, 158–60; patient referral to, 158; physical assessment in, 160–61; practice models for, 164–65; professional credentials for, 157–58; specialized, 157–58; treatment plan in, 163–65
lipid disorders: in adult, clinical evaluation of, 20–26; in children, 103–6; classification of, 10, 11t; complex, 157, 157t; familial, 10–14; genetics of, 10; HIV-associated, 111–20; lifestyle changes and, 23, 25t, 30–40; in metabolic syndrome, 42–55; pharmacotherapy for, 23, 25t, 50–53, 60–80 (*See also specific drugs*); in pregnant patients, 15t, 16, 100–103; renal-associated, 15t, 16, 109–11; risk factors for, 156–57; risk stratification in, 22–26, 22f, 22t; secondary, 14–16, 15t; systematic management of, 156–65; transplant-associated, 16, 109–10, 120–22; treatment goals and strategies, 23, 25t
lipidology, credentials in, 157–58

lipid profiles: determination of, 20–21; fasting, 20–21; nonfasting, 21
lipid tests, advanced, 23–26, 26t, 141–42
LIPID trial, 85–87, 88t, 122, 123t
lipodystrophy, 15t
lipoma, 161
LipoProfile test, 26, 26t, 141–42
lipoprotein(s): apolipoproteins of, 5–6, 6f; in atherogenesis, 1–4, 2f;
  endogenous, metabolism of, 8, 8f; exogenous, absorption and transport of, 6–7, 7f;
  physical characteristics of, 5, 5t; serum, classification of, 20–21, 21t; structure of, 5,
  6f. *See also specific types*
lipoprotein a: in atherogenesis, 1; excess, disorder of, 14; physical characteristics of, 5t;
  as risk marker, 28t, 142
liver disease: ezetimibe and, 70; fibric acid derivatives and, 74; lipid abnormalities with,
  15t, 16; nicotinic acid and, 77–78, 150; statins and, 65, 146–47
Lofibra. *See* fenofibrate
Long-Term Intervention with Pravastatin in Ischaemic Disease (LIPID), 85–87, 88t,
  122, 123t
lopinavir, and statin use, 118t
lovastatin, 60–67; administration of, 67, 144; adverse effects of, 64–66, 145; and
  apolipoprotein B, 143; dosing of, 63f, 64t, 67, 144, 149t; drug interactions of, 66–67,
  68t–69t, 118t, 145; effects on lipid parameters, 62t, 149t; in elderly patients, 123t;
  extended-release, 63, 67; for HIV-associated dyslipidemia, 116, 117t–118t;
  mechanism of action, 61–62, 61f; monitoring with, 146–47; in pediatric patients,
  107t; pharmacokinetics/pharmacodynamics of, 62–64, 63t; in pregnancy, avoidance
  of, 101t, 102; primary prevention trial of, 86; renal considerations with, 112t; for
  transplant-related dyslipidemia, 121t
Lovaza, 79–80. *See also* omega-3 fatty acids
low-density lipoprotein (LDL): advanced tests for, 23–26, 26t; in atherogenesis, 1–4, 2f;
  diet and, 31–36, 54; oxidation of, 2, 2f; pharmacotherapy and, 52–53, 60–80
  (*See also specific drugs*); physical characteristics of, 5, 5t; in pregnant patients, 100;
  serum levels of, 5, 20–21, 21t; therapeutic lifestyle changes and, 30–31;
  treatment goals for, 23, 25t, 84–85
Lp(a). *See* lipoprotein a
LPL deficiency, familial, 12
LRP (LDL-related protein), 7

M

macrophage(s): in atherogenesis, 3–4; in metabolic syndrome, 45
macrophage colony stimulating factor, 3
Management of Elevated Cholesterol in the Primary Prevention Group of Adult
  Japanese (MEGA) trial, 91, 93t
matrix metalloproteinases (MMPs), in atherogenesis, 2f, 3–4
Measuring Effects on Intima-Media Thickness: An Evaluation of Rosuvastatin
  (METEOR) trial, 94, 95t

INDEX

MEGA trial, 91, 93t
Meridia (sibutramine), 54
metabolic disorders: as cause of lipid disorders, 15t, 16; pathophysiology of, 44–48, 45f
metabolic syndrome, 36, 38, 42–55; AHA-NHLBI criteria for, 42, 43t, 49–50; definitions of, 42–43, 43t; ethnicity and, 43–44; gender and, 43–44; identification and management of, 49–55; IDF criteria for, 42, 43t; non-pharmacological therapy for, 50; pharmacotherapy for, 50–55; prevalence data on, 43–44; progression in, prevention of, 50–52; stepwise approach to, 50, 51t–52t; WHO criteria for, 42, 43t
metabolism, of cholesterol and lipids, 5–9
METEOR trial, 94, 95t
micelles, 6–7, 7f
MIRACL trial, 109, 123t, 124
mixed-population trials, 87–91, 90t
MMPs (matrix metalloproteinases), in atherogenesis, 2f, 3–4
monocyte chemotactic protein-1, 3
monounsaturated fats, 33–34
motivational interviewing, 162–63
muscle-related toxicity: coenzyme Q10 for, 154; of fibric acid derivatives, 74; of statins, 65, 66t, 85, 85t, 94, 144–46, 144t
myalgia: coenzyme Q10 for, 154; fibric acid derivatives and, 74; statins and, 65, 66t, 144–46, 144t
mycophenolate mofetil, and dyslipidemia, 119t, 120
myelomatosis, 15t
myocardial infarction: lipid-lowering therapy in, 93t, 94, 106–9, 108t; risk factors for, 31, 32t
Myocardial Ischemia Reduction with Aggressive Cholesterol Lowering (MIRACL) trial, 109, 124
myopathy: coenzyme Q10 for, 154; fibric acid derivatives and, 74; statins and, 65, 66t, 94, 144–46, 144t, 145
myositis, statins and, 144–46

N

National Cholesterol Education Program (NCEP): clinical guidelines in, 84; diet in, 14, 32–36; metabolic syndrome definition in, 42, 43t; pediatric guidelines in, 103–4; risk factors in, 156–57; risk stratification in, 22–26; therapeutic lifestyle changes in, 30–40
National Health and Nutrition Examination Survey (NHANES) III, 44, 44t
National Heart Lung and Blood Institute (NHLBI), 42, 43t, 49–50, 160
National Kidney Foundation, 75, 85, 109–10
National Lipid Association: on fibric acid derivative safety, 75, 76t, 85; lipid clinic resources of, 165; on omega-3 fatty acid safety, 103; patient education resources of, 160; safety task forces of, 85; on statin safety, 65, 66t, 85
National Provider Identification, 163

National Registry of Myocardial Infarction, 106
NCEP. *See* National Cholesterol Education Program
nelfinavir, and statin use, 118*t*
nephrotic syndrome, 15*t*, 16
nevirapine: and dyslipidemia, 114–16; and statin use, 118*t*
NHLBI. *See* National Heart Lung and Blood Institute
niacin. *See* nicotinic acid
Niaspan. *See* nicotinic acid
nicotine. *See* smoking
nicotinic acid, 53, 60, 76–78, 150–52; administration of, 78; adverse effects of, 77–78, 150–51; in diabetic patient, 78, 151–52; dosage forms of, 76; dosing of, 77*t*, 78, 151; drug interactions of, 78; effect on lipid parameters, 77*t*, 150; extended-release, 76–78, 77*t*, 150–51; for HIV-associated dyslipidemia, 117, 117*t*; immediate-release, 76, 78, 150–51; mechanism of action, 76–77; monitoring with, 151–52; in pediatric patients, 106, 108*t*; pharmacokinetics/pharmacodynamics of, 77; in pregnant patients, 101*t*; for renal-associated dyslipidemia, 110*t*, 111; renal considerations with, 113*t*; slow-release, 76, 78, 150; sustained-release, 150
Niemann-Pick C1-like 1 transporter, 6–7
nitric oxide, in atherogenesis, 2–3, 2*f*
NNRTIs. *See* nonnucleoside reverse transcriptase inhibitors
nonesterified fatty acids (NEFAs), 45–48
nonfasting lipid profile, 21
nonnucleoside reverse transcriptase inhibitors (NNRTIs):
  and dyslipidemia, 111–16, 115*t*, 117*t*; and statin use, 116, 118*t*
nonprescription products, 152–54
Norvir (ritonavir), and dyslipidemia, 114–16, 115*t*
NRTIs (nucleoside reverse transcriptase inhibitors), and dyslipidemia, 111–16, 115*t*
nuclear factor kappa B, 3–4
nuclear magnetic resonance (NMR) spectroscopy, 26, 141–42, 161
nucleoside reverse transcriptase inhibitors (NRTIs), and dyslipidemia, 111–16, 115*t*
Nurses Health Study, 46, 143
nutritional labeling, 34
nutritional therapy, 31–36
nuts, 33–36, 36*t*

O

obesity, 38–39; as cause of lipid disorders, 15*t*, 16; definition of, 38*t*; epidemic of, 38, 44; in metabolic syndrome, 38, 42–55, 43*t*; non-pharmacological treatment of, 55; pharmacotherapy for, 54–55; prevalence, in United States, 44, 44*t*
omega-3 fatty acids, 33, 35–36; administration of, 80, 152; adverse effects of, 79, 152; dietary sources of, 33, 35–36, 37*t*, 152; dosing of, 80, 152; drug interactions of, 80; effect on lipid parameters, 79*t*; for HIV-associated dyslipidemia, 116–17, 117*t*; mechanism of action, 79; National Lipid Association on, 103; OTC availability of,

80, 152; in pediatric patients, 108t; pharmacokinetics/pharmacodynamics of, 79; in pregnant patients, 101t, 103; for renal-associated dyslipidemia, 110t, 111; renal considerations with, 113t; supplementation, 36, 60, 79–80, 152; for transplant-related dyslipidemia, 122

omega-6 fatty acids, 33, 35–36, 36t

1%/2% rule, 122

organ transplantation: dyslipidemia with, 16, 109–10, 120–22; immunosuppression and cardiovascular risk in, 119t, 120

orlistat, 54

overweight: definition of, 38t; prevalence, in United States, 44, 44t

P

PAI-I (plasminogen activator inhibitor-I), 2–3, 46, 54, 55

pain level, evaluation of, 161

pancreatitis, 12, 23, 100

pathophysiology, 1–16

patient adherence, 162–63, 163t

patient education, 160

patient evaluation, in lipid clinic service, 158–60

pearls, clinical practice, 139–54

pediatric patients, 103–6. *See also* children, dyslipidemia in

pedometers, 159

peroxisome proliferator activated receptor α (PPAR-α), 74

pharmaceutical care, definition of, 156

pharmaceutical care treatment plan, 163–65; CPT coding in, 163–64; documentation of, 163

pharmacist: and patient adherence, 162–63; role of, 156; systematic management by, 156–65

pharmacotherapy: for acute coronary syndrome, 93t, 94, 106–9, 108t; adherence to, 162–63, 163t; for Asian patients, 91, 93t; clinical trials of, 84–94; for diabetic patients, 50–53, 91, 92t; for elderly patients, 91, 92t, 122–24, 123t, 124; for HIV-associated dyslipidemia, 116–20, 117t, 118t; for hypertension, 53; for lipid disorders, 23, 25t, 50–53, 60–80 (*See also specific drugs and applications*); for metabolic syndrome, 50–55; for obesity, 54–55; for pediatric patients, 103–6, 105t, 107t–108t, 146–48; for pregnant patients, 101t, 102–3; for renal-associated dyslipidemia, 110–11, 110t; for transplant-associated dyslipidemia, 120–22

phentermine, 54

phospholipid(s), 4–5; biological functions of, 4–5; carrier proteins for, 5 (*See also* lipoprotein); intestinal absorption and transport of, 7, 7f; reverse transport of, 9, 9f; synthesis of, 4

physical activity, 30, 36–37

physical assessment, 160–61

pioglitazone, and lipid disorders, 16

plant stanols/sterols: in clinical practice, 149–50; dietary guidelines on, 34–35, 35*t*, 149–50; in pediatric patients, 106, 108*t*;in pregnant patients, 101*t*, 103; renal considerations with, 113*t*

plasma, lipemic, 12

plasminogen activator inhibitor-I (PAI-I), 2–3, 46, 54, 55

polycystic ovary syndrome, 15*t*, 16

polygenic hypercholesterolemia, 11–12, 11*t*

polyunsaturated fats, 33–34

porphyria, 15*t*

PPAR-α (peroxisome proliferator activated receptor α), 74

pravastatin, 60–67; for acute coronary syndrome, 93*t*, 94, 106, 108*t*; administration of, 67, 144; adverse effects of, 64–66, 145; and apolipoprotein B, 143; in Asian patients, 93*t*; and atheroma burden, 95*t*; dosing of, 63*f*, 64*t*, 67, 144, 149*t*; drug interactions of, 66–67, 68*t*–69*t*, 118*t*, 145; effects on lipid parameters, 62*t*, 149*t*; in elderly patients, 91, 92*t*, 123*t*; for HIV-associated dyslipidemia, 117*t*, 118*t*; mechanism of action, 61–62, 61*f*; mixed-population trial of, 90*t*; monitoring with, 146–47; in pediatric patients, 105–6, 107*t*; pharmacokinetics/pharmacodynamics of, 62–64, 63*t*; in pregnancy, avoidance of, 101*t*, 102; primary prevention trial of, 86*t*; renal considerations with, 112*t*; secondary prevention trials of, 85–87, 88*t*; for transplant-related dyslipidemia, 121*t*

Pravastatin or Atorvastatin Evaluation and Infection Therapy—Thrombolysis in Myocardial Infarction (PROVE-IT TIMI) trial, 93*t*, 94, 106, 108*t*

prednisone, and dyslipidemia, 119*t*, 120

pregnancy, dyslipidemia in, 15*t*, 16, 100–103; bile acid sequestrants for, 101*t*, 102; cholesterol absorption inhibitor for, 101*t*, 102; fibric acid derivatives for, 101*t*, 103; statins for, 101*t*, 102; treatment considerations in, 101*t*, 102

prevention: primary, 30, 85, 86*t*, 156–57; secondary, 85–87, 88*t*–89*t*

Preventive Cardiovascular Nurse's Association, 160

primary biliary cirrhosis, 16

primary prevention, 30, 85, 86*t*, 156–57

Prospective Study of Pravastatin in the Elderly at Risk (PROSPER), 91, 92*t*, 123*t*

prostacyclin (PGI$_2$), in atherosclerosis, 3

prostanoids, in atherosclerosis, 3

protease inhibitors: and dyslipidemia, 15*t*, 16, 100, 111–20, 115*t*; and statin use, 67, 69*t*, 116, 118*t*

protein, diet high in, 29*t*, 38–39

proteinuria, statins and, 65–66

proteoglycans, in atherogenesis, 1–2

PROVE-IT TIMI trial, 93*t*, 94, 106, 108*t*, 109

R

raltegravir, and statin use, 117–19

Reductil (sibutramine), 54

red yeast rice, 154

repaglinide, interaction with fibric acid derivatives, 74–75
retinoids, and lipid disorders, 15t, 16
Retrovir (zidovudine), and dyslipidemia, 114–16
Reversal of Atherosclerosis with Aggressive Lipid Lowering (REVERSAL) trial, 94, 95t
reverse cholesterol transport, 9, 9f
Reyataz. *See* atazanavir
rhabdomyolysis: fibric acid derivatives and, 74; statins and, 65, 66t, 144–46, 144t
rice, red yeast, 154
rimonabant, 54–55
risk factor clusters, 42
risk factors, for lipid disorders, 156–57
risk markers: clinical trials assessing, 94, 95t; emerging, 26, 27t–28t, 50, 139–43
risk stratification: for adult patients, 22–26, 22f, 24t; for pediatric patients, 103, 104t
ritonavir, and dyslipidemia, 114–16, 115t
rosiglitazone, and lipid disorders, 15t, 16
rosuvastatin, 60–67; administration of, 67, 144; adverse effects of, 64–66, 145; and atheroma burden, 94, 95t; dosing of, 63f, 64t, 67, 144, 149t; drug interactions of, 66–67, 68t–69t, 75, 118t, 145; effects on lipid parameters, 62, 62t, 149t; for HIV-associated dyslipidemia, 117t, 118t; mechanism of action, 61–62, 61f; monitoring with, 146–47; in pediatric patients, 107t; pharmacokinetics/pharmacodynamics of, 62–64, 63t; in pregnancy, avoidance of, 101t, 102; for transplant-related dyslipidemia, 120–21, 121t

S

4S (clinical trial), 85–87, 88t, 122, 123t
SAGE trial, 123t
St. John's wort, statin interaction with, 145
salicylates, allicin (garlic) and, 154
saquinavir, and statin use, 118t
saturated fats, 33, 34t, 50
Scandinavian Simvastatin Survival Study (4S), 85–87, 88t, 122, 123t
SEARCH trial, 109
secondary lipid disorders, 14–16, 15t
secondary prevention, 85–87, 88t–89t
segmented gradient gel electrophoresis, 141
Segrest, Jere, 141
sibutramine, 54
Sibutramine Trial in Obesity Reduction and Maintenance (STORM), 54
simvastatin, 53, 60–67; for acute coronary syndrome, 93t, 94, 108t, 109; administration of, 67, 144; adverse effects of, 64–66, 94, 145; dosing of, 63f, 64t, 67, 144, 149t; drug interactions of, 66–67, 68t–69t, 75, 118t, 145; effects on lipid parameters, 62t, 149t; in elderly patients, 123t; for HIV-associated dyslipidemia, 116, 117t–118t; mechanism of action, 61–62, 61f; mixed-population trial of, 90t; monitoring with, 146–47; in pediatric patients, 107t; pharmacokinetics/pharmacodynamics of,

62–64, 63t; in pregnancy, avoidance of, 101t, 102; renal considerations with, 112t; secondary prevention trials of, 85–87, 88t, 89t

sirolimus, and dyslipidemia, 119t, 120

skin, evaluation of, 161

smoking: and blood pressure measurement, 160; cessation of, 30, 37–38; status, assessment of, 159

solid organ transplantation: dyslipidemia with, 16, 109–10, 120–22; immunosuppression and cardiovascular risk in, 119t, 120

South Beach diet, 38

SPARCL trial, 87, 89t

specialized lipid clinic services, 157–58

special populations, 100–124; acute coronary patients, 93t, 94, 105–9; Asian patients, 91, 93t; clinical trials in, 91–94, 92t–93t; diabetic patients, 91, 92t; elderly patients, 91, 122–24; HIV patients, 111–20; kidney patients, 109–11; pediatric patients, 103–6; pregnant patients, 100–103

stanols, plant: in clinical practice, 149–50; dietary guidelines on, 34–35, 35t, 149–50; in pediatric patients, 106, 108t; in pregnant patients, 101t, 103; renal considerations with, 113t

statins, 60–67, 144–47; for acute coronary syndrome, 93t, 94, 106–9, 108t; administration of, 67, 144; adverse effects of, 64–66, 145; antiretroviral (HIV) drugs and, 67, 69t, 117–19, 118t; and apolipoprotein B, 143; in Asian patients, 91, 93t; and atheroma burden, 94, 95t; class effect of, 62; in diabetic patients, 50–53, 91; dosing of, 63f, 64t, 67, 144, 149, 149t; drug interactions of, 66–67, 68t–69t, 71, 74–75, 145; effects on lipid parameters, 62t, 149, 149t; in elderly patients, 91, 92t, 122–24, 123t; extended-release, 63, 67; fibric acid derivatives with, 147; hepatotoxicity of, 65, 146–47; for HIV-associated dyslipidemia, 116–20, 117t, 118t; immunosuppressant agents and, 120–21, 121t; kidney toxicity of, 65–66; mechanism of action, 61–62, 61f; for metabolic syndrome, 50–53; mixed-population trials of, 87–91, 90t; monitoring with, 146–47; mortality curve with, 146; muscle-related toxicity of, 65, 66t, 85, 85t, 94, 144–46, 144t; National Lipid Association on, 65, 66t, 85; in pediatric patients, 105–6, 107t, 146–47; pharmacokinetics/pharmacodynamics of, 62–64, 63t; pleiotropic properties of, 61–62, 106; in pregnancy, avoidance of, 101t, 102; primary prevention trials of, 85, 86t; for renal-associated dyslipidemia, 110–11, 110t; renal considerations with, 112t; secondary prevention trials of, 85–87, 88t–89t; special-population trials of, 91–94, 92t–93t; for transplant-associated dyslipidemia, 120–21, 121t

stavudine, and dyslipidemia, 114–16, 115t

sterol precursors, 4

sterols, plant: in clinical practice, 149–50; dietary guidelines on, 34–35, 35t, 149–50; in pediatric patients, 106, 108t; in pregnant patients, 101t, 103; renal considerations with, 113t

STOP-NIDDM trial, 52

Stroke Prevention by Aggressive Reduction in Cholesterol Levels (SPARCL) trial, 87, 89t

studies, landmark, 84–94. *See also* clinical trials

Study Assessing Goals in the Elderly (SAGE) trial, 123t

Study to Evaluate the Effect of Rosuvastatin on Intravascular Ultrasound-Derived Coronary Atheroma Burden (ASTEROID), 94, 95t
Sugar Busters diet, 39t
Sustiva. *See* efavirenz
systematic management, of lipid disorders, 156–65
systemic lupus erythematosus, 15t
systolic blood pressure, 159, 160

T

tacrolimus, and dyslipidemia, 119t, 120
Tangier disease, 13
tendon xanthomas, 10
theophylline, bile acid sequestrants with, 148
therapeutic lifestyle changes (TLCs), 23, 25t, 30–40; adherence to, 162; approaches to, 30; in diabetes mellitus, 50–52; in elderly patients, 124; in HIV-associated dyslipidemia, 114, 115t; in hypertension, 53; in metabolic syndrome, 50–53; multidisciplinary approach to, 30; patient-specific plan for, 30; in pediatric patients, 105; steps in, model of, 31f
thiazide diuretics, and lipid disorders, 15t
thiazolidinediones, and lipid disorders, 16
thromboxane (TXA_2), in atherosclerosis, 3
tipranavir, and statin use, 118t
tissue-type plasminogen activator (t-PA), 2–3
TLC diet, 32–36, 33t
TLCs. *See* therapeutic lifestyle changes
TNT trial, 52, 87, 88t
toll-like receptors, 3
t-PA (tissue-type plasminogen activator), 2–3
trans fats, 33–34, 50
transplantation, solid organ: dyslipidemia with, 16, 109–10, 120–22; immunosuppression and cardiovascular risk in, 119t, 120
transport, of cholesterol and lipids, 5–9
Treating to New Targets (TNT) trial, 52, 87, 88t
treatment goals and strategies, 23, 25t
treatment plan, pharmaceutical care, 163–65; CPT coding in, 163–64; documentation of, 163
TriCor. *See* fenofibrate
triglyceride(s), 4–5; advanced tests for, 23–26, 26t; carrier proteins for, 5 (*See also* lipoprotein); diet and, 31–36; elevated levels of. *See* hypertriglyceridemia; endogenous, metabolism of, 8, 8f; insulin resistance and, 48f, 46–48; intestinal absorption and transport of, 7, 7f; pharmacotherapy and, 52–53, 60–80 (*See also specific drugs*); in pregnant patients, 100; serum levels of, 5, 20–21, 21t; synthesis of, 4; treatment goals for, 23
tumor necrosis factor α, 45

U

ubiquinone, in muscle symptoms, 154
ultracentrifugation, vertical density gradient, 141
ultrasound: of atheroma burden, 94, 95t; carotid B-mode, 140
unsaturated fats, 33–34, 34t, 50

V

VA-HIT trial, 123t
VAP cholesterol test, 25–26, 26t
vascular cell adhesion molecule-1, 3
vascular disease, measurement of, 139–41
verapamil, statin interaction with, 69t
vertical auto profile (VAP), 25–26, 26t
vertical density gradient ultracentrifugation, 141
very low-density lipoprotein (VLDL): in atherogenesis, 1; insulin resistance and, 46–48, 48f; physical characteristics of, 5, 5t; in pregnancy, 100
Veterans Affairs High-Density Lipoprotein Cholesterol Intervention Trial (VA-HIT), 123t
Veterans Affairs model, of lipid clinics, 164–65
Viramune. *See* nevirapine
vital signs, assessment of, 160–61
vitamin(s), 153
vitamin C, 153
vitamin E, 153
vitamin K, bile acid sequestrants and, 72
VLDL. *See* very low-density lipoprotein
voriconazole, statin interaction with, 69t

W

waist circumference, 36, 46, 47f, 158, 159
warfarin, interactions of: with allicin (garlic), 154; with bile acid sequestrants, 148; with ezetimibe, 71; with fibric acid derivatives, 74–75; with omega-3 fatty acids, 80
weight, evaluation of, 159, 161
weight gain, 15t, 32, 38
weight loss, 38–39; dieting for, 29t, 38–39, 54; exercise and, 36; pharmacotherapy for, 54–55
West of Scotland Coronary Prevention Study Group (WOSCOPS), 85, 86t
Women's Health Study, 142, 153
World Health Organization (WHO), on metabolic syndrome, 42, 43t
WOSCOPS trial, 85, 86t

X

xanthelasmas, 10
xanthoma(s): assessment for, 161; eruptive, 12; planar palmar, 13; tendon, 10
Xenical (orlistat), 54

Z

Zerit (stavudine), and dyslipidemia, 114–16, 115*t*
Zetia. *See* ezetimibe
zidovudine, and dyslipidemia, 114–16
Zone diet, 38, 39*t*